PACEM IN MARIBUS

EDITED BY

Elisabeth Mann Borgese

DODD, MEAD & COMPANY
NEW YORK

ISBN: 0-396-06417-5
Library of Congress Catalog Card Number: 72-3140

Printed in the United States of America
by The Cornwall Press, Inc., Cornwall, N.Y.

Contents

Preface

Man thinks himself superior to animals. Although he began his evolution from the same genealogical tree, his brain has developed enormously in comparison with theirs and he has directed his instinct toward intelligence by thinking, imagining, and above all by materializing his thoughts, and definitely and irreversibly becoming engaged in what he names "progress." Undoubtedly he has gained certain material advantages, but these advantages have begun to reflect a dangerous and unique situation. More than any other living species, man is now in a position to struggle effectively against the effects of natural selection.

Through science, technology and his creative genius, he is continually compensating for natural events. He modifies his own structure and even the framework in which he lives. He enables his weak to survive; and he multiplies his numbers to an astounding degree. But, in doing this, he is also endangering his own future, by enfeebling the species to such a dangerous extent that one cannot predict if he will be able to continue to invent the stratagems by which he is maintained a viable entity. Altogether such progress—the ability to multiply and struggle against natural selection—gives to the human species for the first time in his life on earth the possibility of collective suicide. A number of destructive tactics could be mentioned here: Atomic energy, a destructive epidemic artificially launched, and quite simply, an excessive degree of global pollution.

When we consider the question of pollution, we must recall the fundamental importance in the worldwide ecological equilibrium of the upper layers of the sea. Here the phytoplankton absorb solar energy and produce nourishment for zooplankton and consequently for all living species of the ocean. More than half of the world's production of oxygen is fabricated by this marine phytoplankton which has the further task of absorbing the excess of carbon dioxide that is produced by animal respiration and, for several generations past, by human industry. It is not difficult to see the important role played by this thin oceanic layer. And yet we annually discharge into it hundreds of thousands of tons of lead, mercury, detergents, and various insecticides, not to mention a million tons of petroleum, an amount equal to about one per cent of world production. In all these ways, the phytoplankton is directly threatened by pollution. If it were to disappear, life in the sea would become extinct and possibly after a brief interval much of terrestrial life, suffocated both by a lack of oxygen and an excess of carbon dioxide, would also cease to exist.

Phenomena like these, and their corollaries, are so complex and still so little understood that it is not at all certain that scientists will be able to find remedies before these ills turn into absolute catastrophes. What is certain is that the only chance of human survival lies in systematic and global study of these problems; and that to carry out such study we must knock down all barriers, whether they be racial, social, financial, ethnic or political. Only an understanding on the part of all the peoples of the globe, of all their organizations and societies, will assure positive results—assuming that we still have time.

The sea must continue to play a vital role for the whole human race, even though today all the old priorities that existed until recently have disappeared. It is not so much a question of nourishing humanity. But rather, we must guard against preventing the sea from carrying out its irreplaceable contribution to our respiration. In short, what we take out of the sea is no longer as important as what we do not put into it.

Here is the great thesis to the defense of which *Pacem in Maribus* can contribute with all consciousness, conscience, and energy.

JACQUES PICCARD
(Translated by John Wilkinson)

Introduction

The Marine Revolution is upon us, and now must take its place on the long list of great disjunctures that have marked human history—the political, industrial, socio-economic revolutions of the past, the technological and biological revolutions of the present. The Marine Revolution partakes of all of these and adds a new dimension. Potentially it is a revolution in international relations.

The great sea change stems immediately from the rapidly expanding and intensifying industrialization of the oceans. Scientific and technological breakthroughs have opened the hidden depths, and in the process they have raised a host of ecological issues related to the increasingly acute concern for the total human environment. As man moves for the first time to exploit territory traditionally regarded as no-man's land beyond sovereign claim, he poses grave new problems of development and disarmament and brings new stress to the fragile structure of international relations.

There are already many ominous signs that the Marine Revolution could turn out to be predominantly destructive. In important ways it is without precedent: starting from a far more advanced stage than earlier industrial revolutions this impending transformation allows no time to adjust to change; and it takes place at the confluence of pollution from air, land, and water in a medium that magnifies the effects of miscalculation. On the basis of present trends reputable scientists now predict that the oceans may be dead of man-made pollution before the end of the century.

Bereft of this essential reservoir of life, the earth might finally become unable to sustain the marauding human race.

Yet, no one can seriously propose that industrializations of the oceans be halted. A "zero-growth economy" for the seas is the most utopian of all utopias—and worse still, it is a rich man's dream that would become a nightmare for the majority of peoples whose survival requires full development of the world's resources. Luddism did not work on land. It will not work under water.

The realistic alternative is to harness and rationally direct the forces of the Marine Revolution, minimizing its destructive side effects. Then the oceans can be bountiful.

Food production, which actually is decreasing as a consequence of pollution on the one hand and heedless overfishing on the other, could instead be increased fourfold, even sixfold, during the next thirty years. Oil production might well increase at about the same rate. Lockheed Offshore Petroleum Services (Canada) recently conducted tests of an underwater production system designed to operate in 1,200 feet of water. The Institute Français du Petrole uses a system that permits complete remote control of drilling and coring. An undersea storage tank and tanker loading device system is being installed in the Bay of Biscay. This, according to *Ocean Industry* (October, 1970), is an economic replacement for the common, but expensive method of installing a network of pipe lines loading from the field to onshore storage facilities, then returning to an offshore loading terminal designed to service tankers.

Advancing technology will reclaim an increasing proportion of the mineral treasure trove on the ocean floor. A Japanese group has developed a mechanical system to mine manganese nodules from the deep of the ocean floor. It consists essentially of a continuous loop of cable to which dredge buckets are attached at regular intervals. Design calculations indicate that a full scale prototype of this continuous line bucket dredging system (CLB) should be able to produce about five thousand tons of the nodules per day from up to fifteen thousand feet of water for a production cost of about two dollars per ton of nodules recovered. The capital investment in such a system, exclusive of the cost of the vessel which would be chartered (as this system can be operated off almost any type of ocean going ship) would be about one million dollars. It seems there are about 1.7 trillion tons of nodules distributed over the deep ocean floor in the Pacific alone. The mining of one per cent of the ocean bottom would satisfy the world's need, at the present level, for manganese, nickel, copper, and cobalt for about fifty years. This transcends the boundaries of the Marine Revolution. It would mean a revolution in the mining industry, worldwide, and would affect, in particular, a number of developing nations whose economy today depends almost entirely on the export of these metals.

For better or for worse, we can expect cities to expand over the oceans, and colonies for work and recreation to come into being deep down below. The oceans are an essential part of any system providing improved means of weather forecast and control; communications and transport on and below the surface are destined to grow in volume, density, and speed. The Marine Revolution has brought into a new focus the basic issues inherent in the technological and biological transformations that characterize our age, and this in turn demands consideration of their impact on democratic institutions and international relations. The oceans have come to pose a problem too serious, and too diverse, to be left to oceanographers, a problem that is as interdisciplinary as it is transnational, positioning ecology as the foundation of law, pointing towards a transcendence of the concept of sovereignty and territorial boundaries, in a medium in which these simply cannot apply, postulating the existence of a common heritage of mankind as the basis of a new relationship among all nations, rich and poor, and of development, no longer dependent on "aid" but on the sharing of that common heritage, proposing the establishment of a "peace-system" of interlocked scientific and industrial enterprises which—for the first time in fifty years of frustrating negotiations, raise a real possibility for the abandonment of the arms race.

Thanks to the bold, imaginative initiative of the government of Malta, these issues were joined before the United Nations late in 1967. A forty-two-nation Seabed Committee was given a mandate to propose to the General Assembly a set of principles to support the legal framework and functional structure of an international ocean regime. Thus the need for action has been recognized, and pioneers on the frontier of evolving political theory can address themselves to a ready forum.

It was in this context that the Center for the Study of Democratic Institutions embarked five years ago on a comprehensive study project, Pacem in Maribus. The Center sponsored five major conferences of international authorities on disarmament, development, maritime law, ocean ecology, and ocean enterprises. Each of these conferences produced a volume of research papers and discussion transcripts. This work culminated in the Pacem in Maribus Convocation in Malta during the summer of 1970 which assembled about 260 political leaders, industrialists, scientists, and fishery experts and brought the strands together on the assumption that the endeavor had reached the point of passage from theoretical research to consideration of the operational questions that confront those who have the authority to take political action.

The Malta Convocation set up a Pacem in Maribus Continuing Committee for Policy Research which now serves as an international board of directors for an autonomous Pacem in Maribus Institute incorporated in the Royal University of Malta. Five new study projects are in course, and a second Convocation was held in Malta in 1971. Pacem in Maribus III is scheduled for the summer of 1972.

This volume presents a selection and condensation of this work until now for the nonspecialized reader. The full material, in six volumes plus a bibliography, has just been published by the University of Malta Press.

In these introductory pages we shall try to give a bird's-eye view of some of the major problems faced during our work and which must be solved if an international ocean regime is to be established. Imaginative solutions to these problems, however, would do more than give rise to an ocean regime. Their applicability may turn out to be wider than the oceans. A successful model for an ocean regime might provide a new pattern for international organization in general.

Since the ecology of the ocean environment must be the basis of the law of the oceans, we shall construct our marine architecture from the foundation and begin with scientific problems, follow with those of the "producers" whose *rational use* of the oceans must be based on science and governed by law, and thus end up with the legal and political issues. The basic theme of our scientists was the "Role of Science and Scientists in an Ocean Regime." It might as well have been the "Role of Science and Scientists in the Political Order," or the "Role of Science and Scientists in the World Order." For what is central and crucial in an ocean regime turns out to be crucial for world order or for the political order in general.

The discussion centered on the problem of freedom of scientific research and investigation, its crisis today, and ways to resolve it in the context of existing or new institutions.

The causes of the crisis are partly contingent, partly intrinsic. Contingent are the restrictions arising from growing nationalism and from the rational and irrational preoccupations of the developing nations and their tendency to make increasing areas of ocean and ocean floor inaccessible to scientific research, whether national or international. If the developing nations were concerned lest the natural resources off their shores might be prospected and subsequently exploited by foreign commercial interests, under the pretext of free scientific exploitation, these fears were not allayed by the discussions.

Attempts were made to draw the line between scientific and commercial exploration, but the line seemed blurred. It is obvious that any scientific exploration yields advantages both to the military and to industry. Guarantees offered to coastal nations, including the participation of experts of the coastal nation in the exploration, the sharing of all data and samples, and the publication of all results, are not attractive to developing nations which often do not have the experts to associate with the foreign expedition, and no use for the results of the exploration. They fear the damages to their fish stocks off their coasts, from seismic explosions and the oil spills from deep-sea drillings, and prefer to protect their interests their own way.

These restrictions are the contingent causes of the crisis. The real cause lies far deeper. The real cause is that freedom of science, as a corollary of the old freedom of the seas, is in crisis like all freedoms in this twentieth century.

To stay, for the moment, with the oceans: as the concept of the freedom of the seas is being superseded by the concept of the common heritage of mankind, so the freedom of science must be re-interpreted and re-embodied in a new conceptual and institutional framework. Part of the trouble lies within science as such. Part, in the relationship between science and society.

The internal aspect of the crisis—or transformation—was dealt with by the Swedish ecologist, Bengt Lundholm. The new science, "integrated ecology," he explained, includes resource management, and it should help society to predict changes. Its two main areas of study are the energy flow and the flow of nutrients or chemical elements through the eco-system. If one wants to study this, one has to study producents, consumants, and destruents, i.e., green algae, human beings, and bacteria. The new idea of this ecology is that these three parts have equal importance in the eco-system. If you want to study their interaction in an eco-system, you must study all parts. "As a consequence, research must be organized in a very special way, and in a way *which partly is interfering with the freedom of science and the scientist.* You must have complete teams of scientists working in the total field of research. The quality must be problem-oriented. You must use your resources—economic resources, especially—to build your cadres . . . *Just as industry and the military must hire and manage its scientists, so must ecology.*"

This rather dramatic restriction of "freedom" in the old sense thus comes *from within science itself.*

As always, outside corresponds to inside. If science, impelled by its internal evolution, must act the way industry and the military do, it is industry and the military that are threatening and undermining the freedom of research from outside. While science and technology have taken on an unprecedented and all-pervasive importance in contemporary society, affecting the life of every citizen, of every living being—while it has become a productive force, a means of production—it is today largely determined by the interests and normative axioms of those who in a given society exercise the real power. This was the thesis of the Swiss philosopher, Arnold Künzli, which he has developed in the chapter "Science as the Social Property of Mankind."

If science is the common heritage and social property of mankind, society has the right and duty to make sure it is not used for purposes opposed to the legitimate requirements of society. This control must be a social control. It must not be allowed to fall into the hands of a national or international bureaucracy or a party apparatus. In other words, it must be a social self-control, in which science and scientists participate in a measurable way to secure an optimum of freedom of scientific research which today can be achieved only in this context.

Most people agree that the old, intergovernmental formula of science administration has run its course—that, reflecting a nineteenth century concept of science as passively observing and describing nature, it has lost its efficacy. This leads, on the one hand, to catastrophic gaps between knowledge and action, as was the case when scientists fully well knew what was happening to the blue whale, but politicians failed to act on that knowledge and brought this marvelous beast to the edge, or past the edge, of extinction. On the other hand, it engenders duplications of effort that degenerate into waste. There are today at least thirteen intergovernmental agencies and fourteen committees at the world level that are dealing with matters of science and technology within the U.N. system, and sometimes their frames of reference are literally the same.

Experts point out that these agencies are essentially vertical organizations, arising from the needs of government activities as seen at the end of the last century. There is no concept of a multidisciplinary horizontal systems approach to activities. It seems unlikely that governments will solve this problem in the near future. It can be solved either by a total crisis or by a group of people putting forward really new ideas.

The new concept of science as not only observing, but changing nature, must be embodied in a link between research and action. This requires institutional innovation to enable scientists to participate in the decision-making processes of planning and government.

How can this participation of scientists be organized and articulated? Appointments of scientists to full-time administrative and executive positions are not easy to fill. They alienate the scientist from his science. Running for elective political office effects the wrong kind of "natural selection" among scientists. It is not the science-minded but the go-getter, often unsuccessful in his own field, who will run. Scientists might try to act on governments through the pressure of public opinion, but this presupposes a long process of education, and communication of the highly complex matters of science and science policy through the mass media would in any case remain a difficult, not to say, insoluble, problem.

Two models for a new scientific policy have been produced by the Center for the Study of Democratic Institutions. One proposes the "constitutionalization of science," its professionalization, internal democratization, and organization into a sort of Parliament of Science. It would bring politics into science. The other is a structural part of *The Ocean Regime.* It would bring science into politics. It proposes a Chamber of Scientists which, without presupposing fundamental changes in the internal organization of science or interfering with its autonomy, would provide a forum where representatives of scientific organizations, international or national, public or private, could discuss problems of common concern among themselves *and* with the political decision-makers. Any decision touching on science or science policy would have to be passed both by the political House or Chamber and by the Chamber of Scientists, or, if they fail to take action, by a majority of delegates of both chambers in joint session. This provision would enable scientists to participate effectively in decision-making, without some of the drawbacks of other solutions. Since participation in the Chamber of Scientists would by no means be a full-time job, participants would not be alienated from their science; nor would it imply political campaigning since, presumably, participants would be nominated or delegated by their own scientific organizations or institutions.

While this problem is of general importance, it is acute and urgent as far as the management of the ocean environment is concerned.

As matters stand today, it is quite clear that ocean policy is not derived from ocean science. Ocean policy reflects national political interests which, in turn, are based on economic, security, social, and philosophical grounds, and science plays only a relatively small part in ocean policy. This may have been all right in the past when ocean use was not dependent on scientific knowledge, but we are reaching a point when scientific knowledge of the ocean is going to be increasingly important in the use of the ocean and its resources, because we are moving from a phase of exploitation to a phase of rational use. So science must become a major element in decision-making in the oceans.

The management of nonliving resources and that of living resources pose totally different problems calling for totally different solutions. Fishing can be carried out by relatively small enterprises, with small capital investment and a simple, even primitive, technology. The extraction of minerals and oil, on the other hand, calls for huge investments, highly sophisticated technologies, and, accordingly, large enterprises operating on long-term leases and licenses guaranteeing fair returns on their investments.

The large-scale extraction of nonliving resources constitutes a *new* use of the ocean environment and calls for *new* regulation. Fishing is an *ancient* use, covered by a variety of rules, local, regional, and international. To scrap these, whether or not they work, and to institute *ex novo* a universal set of rules for the global management of fisheries would be uneconomical, to put it mildly. To subject fisheries to the *same* rules as mining and oil industries would be nonsensical. This, however, does not mean that the fisheries must remain outside an ocean regime; that they should be asked to carry on as of old, heedless of newcomers and new users, of the advent of new civilizations and the passage of centuries.

Fishing depends on the detection and conservation of stocks, which depends on the ocean sciences.

Fishing interferes, actively and passively, with the laying of cables, the operations of mining and drilling, and the military uses of the oceans—as long as these exist.

The fisheries are the first victims of pollution.

Thus ways must be found for the fisheries to deal with the Ocean Regime and for the Ocean Regime to deal with fisheries.

The relationship with a regime for all uses of the oceans raises four basic problems:

—the freedom of fishing and the property status of living resources

—the relation between regional and global international fishing arrangements

—the relationship between developed and developing fishing nations

—the kind and amount of planning required for the world's fisheries.

No one doubts any longer that the right to fish and the freedom of access to fish stocks must be regulated by the requirements of conservation. Unlimited freedom of competitive fishing first raises costs. The same amount of fish could be caught with a fraction of the manpower and shipping volume actually employed. Secondly it also leads to the extinction of species after species, of fishery after fishery. Obviously it is another one of those freedoms to kill freedom. To save this freedom by limiting it is in fact the gist of the two-score-and-odd major and minor fishery treaties and conventions regulating the catch of different stocks or in different regions. These treaties, while solving some problems for some time, leave many others unsolved. Overfishing is still a burning problem in many cases, and new fishing nations and new enterprises, not included in the old conventions, are knocking at the doors or crashing the gates. The question, quite succinctly, is whether a *limited economy*, as the world's fisheries have become, can maintain the *unlimited* "common property" concept for its resources, implying theoretically free but practically much embattled access, theoretically for all but practically for those who have the power to prevail—or whether the old concept of "common property" must be replaced by that of "common heritage" with its precise legal implications of non-appropriability, conservation for posterity, shared management and shared benefits.

All odds point in the latter direction, and while international lawyers keep pointing out that international law provides nowhere any basis for the collection of rents or royalties on fishery produce —which would be one aspect of applying the "common heritage" concept—economists and fishery experts have long since begun to give serious consideration to this innovation. Controls over access must be established, and they can be established by taxes, license fees, auction or lease of rights, or other similar means. Society would benefit by receiving a share, or the full amount, of the surplus profit.

Although regional fishery arrangements have proved successful in a number of cases and, where they have been successful they certainly should be maintained, there is today no longer any sort of consensus among fishery experts as to whether this is where the future lies. Historically, such arrangements began to be made at the beginning of this century, when nations first became aware of the danger of overfishing and the need for conservation measures. Today, with the great fishing nations dominating the world's seas and oceans, these solutions seem somewhat antiquated. In ecological terms, there are many problems which defy solution at the regional level. Many species travel farther than regional arrangements can reach. Many kinds of pollution have global dimensions. In political terms, or in ethical terms, regional arrangements fail to apply or to advance the application of the concept of the common heritage of mankind. The coastal states, or the states party to the regional convention, get all. Nothing goes to mankind as a whole; while there is no reason why the interests of the coastal state should prevail over those of the world community any more than those of the technologically advanced nations. The claims of the coastal nations are not sacrosanct either.

In spite of the apparent advantages to member states, however, the regional conventions have, on the whole, surprisingly small numbers of signatories. Four reasons have been adduced: regional arrangements are hard to apply and do not make much sense so long as the limits of the territorial waters remain undefined by international law; they give too many advantages to coastal states; they give too little advantage to coastal states; and they provide for compulsory settlement of disputes, which many states are not ready to accept.

The logical compromise between "regionalists" and "globalists" would seem to be to leave fishery *management* to regional and local arrangements wherever this proves practicable, while setting up some machinery for the determination of *standards* and *criteria* which are universally applicable or require action on an inter-regional or world scale.

* * *

The success-story of the rapid economic progress of the developing nations due to their free and easy access to fishery resources and the fishing industry, did not, alas, hold up in the light of the discussions. The spectacular increase in the rate of production of the developing nations turns out to be due, entirely, to the exceptional factor of the Peruvian anchovetta industry. If this were to be deducted, the increase would boil down to less than one percent. Of the larger fishing vessels (over one hundred tons in size) capable of moving into distant waters, something like eighty-one percent are owned by seven states none of which can be listed as developing, as Francis T. Christy pointed out. Forty-nine percent are owned by the Soviet Union, and the balance by Japan, Spain, the United Kingdom, France, Norway, Poland and West Germany. The fishermen of the developing nations are poor. They operate on a subsistence level, with a low margin of profit. Processing and marketing is in the hands of big companies, who operate with foreign capital on the world market. Large-scale industrial ventures, such as the harvesting of Antarctic krill, are inaccessible to the developing nations.

Fish, after all, are a raw material; fisheries are one more component of the post-colonial extraction economy which, during the decades since the end of World War II, has served only to widen the gap between the developed and the developing nations.

Participation, on the part of the developing nations, now takes four different forms: participation in decision-making, especially with regard to access to fishery resources; direct participation in the catch; participation in profit sharing; and participation in joint ventures.

Spokesmen for the developing nations keep pointing out that the first is severely limited by existing conventions and treaties; the second is a slow vehicle of progress, given the lack of technology; the third by itself would not only not solve the problem but might even aggravate it: the receipt of payments without the right to participate as equals has psychologically damaging effects and often serves only to make the few rich richer, in the developing nation itself, while leaving the poor poor.

If the development of fisheries is left to flow through the market mechanism in the traditional fashion, it is very doubtful that the objective of advancing the cause of the developing nations can be met on a short time scale, that is, short enough so that they can benefit in our lifetime.

With the exhaustion of more and more stocks in the northern hemisphere, fisheries, no longer profitable to the industrially developed nations, are migrating south. This aggravates the division between the two hemispheres: the North being capable of the industrial exploitation of nonliving and living resources, and the South fishing as of old—of very old—and fishing for the North.

The best hope for the developing nations, with regard to fisheries as to everything else, is a strong, effective international organization in which they can participate as equals. If the ocean regime were embodied in such an organization, a mechanism could be set up in which fishery experts and operators could determine, in cooperation with the other users of the ocean environment, those standards and criteria that have to be determined at the world level. The management of fisheries resources and the execution of plans should be left to the now existing local and regional arrangements.

Such over-all planning at the world level should include the setting up of priorities with regard such questions as: extraction of mineral resources as against living resources; harvesting of one living resource as against another (stocks); one method of extraction as against another (gear); and development of new technologies as against nondevelopment, where such development might be harmful and interfere with other uses. Such planning must not be topheavy. It must be articulated in such a way that the small operator is involved. It must be done from the grass roots, or otherwise it won't work.

With the scientists and the "fish people" thus fitted into the deliberations of the ocean regime, it seems logical to assume that the participation of enterprises engaged in the extraction of nonliving resources should follow the same pattern.

Among the Pacem in Maribus experts, there was agreement on two points: first, that the role of these enterprises is one of crucial importance to the successful operation of an ocean regime; and, second, that the interests of socialist and capitalist, of public and private or mixed enterprises, are the same, so far as the extraction of minerals from the oceans is concerned.

Opinions divided, and new ideas emerged, in two areas: the relations between enterprises and the regime; and the relations between enterprises and the developing nations.

One school of thought holds that the ocean regime must be embodied in an *inter-national* organization of the traditional kind, i.e., one that deals with nation-states exclusively, and not with nongovernmental and nonsovereign entities such as enterprises. In the United Nations this point of view is strongly advocated by the Soviet Union and by some of the other socialist nations, with the proposition of a "two-tiers" system. That is, the international organization would grant licenses to nations. Nations would then grant them to enterprises. It has been argued that the "two-tiers" system would be simpler, require less machinery and that it would more clearly fasten responsibilities and liabilities where they belong, namely on the States. The other school of thought recommends that licenses be issued by the international regime directly to enterprises. Considering the rapid advance of technological integration, the growing interdependence of the world market, and the evolution of the huge multi-national corporations, one may indeed come to the conclusion that the trend, during the next twenty years, is such that resource exploitation by nation-States will become totally unrealistic. The direct relationship between regime and enterprises would strengthen international law and circumvent many of the obstacles generally placed in the path of international cooperation by consideration of national sovereignty. It would cut down red tape and paper work because, instead of having to deal immediately with about 130 nation-state applicants, the regime would have to deal with the very few technologically and financially equipped companies capable of handling the problem of ocean mining.

It should be noted that both positions have their advocates both East and West, among capitalist and socialist enterprises. Either solution holds advantages. It is even thinkable that no rigid, once-and-for-all decision is necessary, and that both principles could be combined in a flexible way. Certainly there will be applicants who should be granted licenses although they are not nation-states, as, for instance, the European Community. This in itself is a departure from the *status quo*, but the march of history makes such departures both inevitable and justifiable.

It should also be noted that the participation of enterprises in a chamber of industries does not depend in any way on a decision regarding two-tiers versus direct licensing. No matter from whom they get their licenses, the enterprises engaged in the extraction of nonliving ocean resources must have a place in the structure of the regime, where they can deliberate on their own problems and requirements, carry on a dialogue with the other users of the ocean environment, and participate in the decision-making of the political organs. With regard to this last point, there is already a precedent in international organization. The ILO's experience in this area might well be adapted to the needs of an ocean regime.

The interests of the multinational corporations are, in many cases conflicting with those of the developing nations. Neither national governments nor international organizations really have the power to put the necessary constraints on the giant corporations. Discussion of these matters is urgent and essential, but the corporations are not always anxious to air this particular aspect. Conflicts may arise between the corporations and the developing nations during every phase of the production process. This concept has been developed by Robin Murray in Chapter 21.

An alternative to the unrestrained operations of the great, multinational corporations is provided by the joint venture. This may well turn out to be the more promising form of industrial co-operation between developed and developing nations, since it provides for the participation of the developing nation in the exploitation of resources on a flexible scale.

It can be said that the scientists and enterprisers have backed more or less reluctantly into complex legal and political issues raised by their interventions into the ocean depths—and many of the enterprisers, at least, would like to see the deep seas continue as a kind of international no man's land. When the undersea area became of prime interest to the military, however, questions of sovereign rights and jurisdiction rose inescapably to the top of the political agenda of the great powers as they maneuvered uneasily on the cold war's diplomatic front.

After long negotiations, the Geneva Disarmament Committee adopted unanimously a Russian-American Draft-Treaty to keep the ocean floor denuclearized. The draft was approved by the General Assembly and has been ratified by large numbers of nations.

It is fair to say that none of the Pacem in Maribus experts was very happy about the Treaty which prohibits the implanting and emplacement of weapons of mass destruction on the ocean floor and its subsoil beyond a twelve-mile limit from shore. There is no ban on moving objects, whether nuclear submarines or slowly moving unmanned launching platforms. Within the twelve-mile limit, furthermore, anything could be installed by a coastal nation. Inspection and verification is left to nations individually. There is no reference to the internationalization of controls. In case of conflict, parties have the right to withdraw from the Treaty without any provision for compulsory arbitration.

The Treaty is not viable from the point of view of environmental reality: the ocean environment is opaque; inspection is impractical. It is not viable from the point of view of military reality: there is today an irresistible pressure to transfer land-based weapons systems into the oceans. Land-based missile systems are dead, one expert said. Strategy will not depend on them any longer. The sea remains the ultimate and essential element in the strategy of the superpowers.

The Treaty is not viable from the point of view of technical reality: the nonatomic nations today have no capacity to engage in active and independent inspection activities. It is inadequate from the point of view of legal reality: the Treaty subjects fixed installations, juridically, to a different regime than that covering mobile operations; to be effective a treaty covering the ocean floor would have to be complemented with provisions covering superjacent waters. And, finally, it fails from the point of view of political reality: the objectives of developed and developing nations are radically different. The former aim at arms control to relieve the cost and burden of the arms race; the latter demand complete disarmament in the hope of neutralizing one of the advantages the technologically developed nations are not yet disposed to give up.

Thus one may dismiss the whole thing as a public-relations gimmick and agree with one of the experts who said "the only way to stop the arms race is to refuse free public relations to those who want to maintain it." Or one may come to the melancholy conclusion that, had the Treaty been defeated, this would have been worse yet—a symptom of complete hopelessness—than its adoption by the General Assembly.

The other two basic legal questions—interrelated—that came under protracted discussion during all phases of development of the Pacem in Maribus project were: the limits of national jurisdiction and the boundary of the legal continental shelf; and the problems raised by the Nixon proposal of May 23, 1970, subsequently expanded and explained in an American "working paper" introduced in the Seabed Committee in Geneva in August, 1970.

As far as the boundary question is concerned, there is today a general agreement that the Continental Shelf Convention of 1958 must be revised. Most people know, however, that there will be no agreement on a better defined boundary—not for many years to come.

The most diverse criteria are being proposed. A new depth limit: two hundred meters, five hundred meters, 3,500 meters. A horizontal limit: fifty miles, one hundred miles, two hundred miles. Combinations between depth and width: two hundred meters depth or fifty miles out, whichever is farther. Then, there are the advocates of the geological boundary: the continental slope, the continental rise. Find the point where the rock formation characteristic of the continents touches the abyssal plain, typically formed by different rock. The research necessary to draw that boundary would cost a few billion dollars. Concepts no longer valid on the continents, left behind by advancing technology, such as the geological boundary, the strategic boundary, are automatically transferred into the new medium.

None of the boundaries proposed has anything to do with the ecologic reality of the oceans: with the control of pollution, with the conservation of fish stocks, and with currents and waves.

The great maritime nations are under conflicting pressures as to where to draw the boundary. The Navy, honoring the tradition of the Freedom of the Seas, in general wants a narrow territorial sea—leaving it free to operate as close as possible to other shores. Commercial interests—especially oil—are powerfully lobbying in favor of the widest possible claim to the continental shelf.

The developing nations, on the other hand, are also extending their claims over ever wider expanses of territorial sea and continental shelf. This is their chance to enrich themselves, many of them seem to be thinking.

Mainland China has just clamorously joined the Latin American Nations in their claim for a two hundred mile limit for their territorial sea. China vituperates the Russian-American position in favor of a three or twelve-mile limit as "socio-imperialism."

Thus, whether the developed nations claim wide or narrow limits, it is always "imperialism." Which does not prove that the developed nations are not "imperialistic." What it proves, instead, is that the question of territorial boundaries has very little, if anything, to do with modern forms of "imperialism" or exploitation.

As though it were territory or natural resources that the poor, developing nations were lacking. They have them, galore. What they are lacking is capital, technology, and the social infrastructure to make use of their vast territories and their abundant resources. By adding more territory and more resources, they solve nothing. All they do is to extend the surface of their vulnerability, of their exploitability by others.

International law is currently dealing with at least five different types or sets of boundaries simultaneously. There is the boundary of the territorial sea, the boundary for exclusive fishing rights, the boundary for disarmament and arms control on the ocean floor, the boundary for pollution control, and the boundary for the sovereign right to explore and exploit mineral resources. A very, very complex system.

It is easy to predict that the question of the boundaries will not be settled until such time as it will have lost all interest. We must have an international regime in spite of that. What is more: the problems the regime must deal with and solve have nothing to do with national boundaries. Pollution must be controlled on both sides of the national boundaries, no matter where you draw them. It must be controlled in the ocean environment as a whole, or it will not be controlled at all. The same applies to the conservation of fish. Fishery conventions, as a matter of fact, apply across boundaries.

What we are looking for, and what we must determine thus is not so much a *geographic* area beyond the limits of national jurisdiction as a *functional* area beyond the limits of national jurisdiction. In other words: what nations cannot do individually, competitively, such as the control of pollution, they must do internationally, cooperatively.

President Nixon proposed that the United States and all other States should renounce any claim beyond a depth of two hundred meters. The whole area beyond that limit should be recognized as the common heritage of mankind, to which a true international regime should apply.

The "international area" would be divided into two parts: a "trusteeship zone" under the administration and responsibility of the coastal nation, extending from the two hundred meter depth limit to the end of the continental margin; and an "international zone," beyond the continental margin, under international administration. Royalties on a scale to be determined, would be paid to "an international organization" on extraction of minerals within the "trusteeship zone." Such payments could begin at once—as soon as a sufficient number of nations agreed to this proposition, for an "interim period" up until the establishment of the final regime.

This regime would apply to the whole area, starting with the two hundred meter depth limit. It would be composed of three parts: a treaty, subsidiary regulations, and machinery. The treaty itself would contain the basic rules and would start with a set of principles. The regulations envisaged are of two kinds. Some would be annexed to the Treaty. The Authority would have as one of its principal tasks to change and adapt regulations to new technological developments and to provide new regulations for subjects people did not think of.

The machinery would be both regulatory and administrative.

As far as it is regulatory, it would basically apply to the whole area, starting at the two hundred meter depth limit. Administrative machinery, however, would be to some extent divided. Certainly there would be some international supervision even in the trusteeship area, especially with regard to pollution. On the other hand, there would be some national authority exercised in the international zone, for licenses would be issued to states on behalf of industries, and thus states would maintain their national responsibilities in these concession areas.

The international machinery would have four main functions: regulation; supervision of the execution of regulations and, if necessary, bringing parties before an international tribunal; cooperation with existing international organizations; and assistance to developing nations, with assurance of equal access opportunities to all.

The Treaty organization would consist of an Assembly in which all members would have to be represented. A smaller Council would represent both the chief industrial powers and the developing nations in such a way that neither of the two groups could impose its views on the other. Finally, there would be a series of operating Commissions, dealing with separate aspects so that power is divided: different Commissions for regulation and supervision, insuring that boundaries are properly drawn by various countries. The Treaty also would provide for an international maritime Court.

The coastal-nation administrative agency within the trusteeship zone, on the other hand, would have two main functions: to collect and divide royalties between the coastal-state government and the international organization; and to decide who can obtain licenses to work in that area, although these decisions would be subject to the standards and regulations of the international regime.

The American government is to be highly commended for taking the initiative of introducing such an elaborate and precise working document. There can be no doubt that the proposal, coming from one of the great powers, has moved the debate to a level of concreteness it did not have before. Certainly, the document is not perfect, but it is not meant for adoption in its present form. It is a working paper, and even if it were better than it is, it would have to undergo a lengthy process of political bargaining before becoming acceptable to the world community.

Pacem in Maribus experts had a number of criticisms:

The division of ocean space into three zones—national, international, and trusteeship zone—was held to be impractical. It might create more problems than it solves.

The boundaries proposed would be unacceptable to many nations, among them the Latin Americans.

The legal status of the trusteeship zone is ill-defined. It is not clear whether the zone is under national or international jurisdiction.

The concept of trusteeship is ill-defined and mis-applied. "Trust" means that the legal interest rests in one person and the equitable interest in someone else. The concept is completely meaningless unless there are effective provisions for the enforcement of trust. Inasmuch as the trusteeship zone is to be established before there exists an international regime or "trustor," inasmuch as it results from a unilateral action by the coastal nation, the concept of trusteeship is really perverted. It is an *inverted trusteeship*, logically and legally unacceptable. A "trust" furthermore, must not fatten the trustee. It is not clear in the proposal whether or not the royalties which the coastal state would be authorized to collect would be limited to the covering of administrative expenses. Also, the stipulation that the United States tax laws should be amended so as not to discriminate against its nationals operating in the trusteeship zone, may arouse some suspicion. Does it mean that oil companies would not be subject to national taxation in the trusteeship zone? This would be a proposition of international concern.

The trusteeship zone detracts from the common heritage of mankind.

The name itself, "trusteeship zone," is unfortunate, because it evokes associations such as Southwest Africa.

The proposal discriminates in favor of some coastal nations, to the detriment of landlocked and shelf-locked nations.

The proposal would abolish the Moratorium on Exploitation of the Seabed Beyond the Limits of National Jurisdiction, adopted by the General Assembly last year.

The "machinery" does not adequately protect the equal rights and participation of the developing nations. Basically, the proposal does not change the relations between the developing nations and the companies of the developed nations. It leaves coastal states free to explore and exploit as far as their technologies will permit, out to the fuzzy edge of the continental margin. This limit is far out in space and time.

The proposal satisfies the demands of the oil companies for the widest possible national jurisdiction, limited only by a *code* to which the administration of the "trusteeship zone" would be subjected, without any international machinery.

To be viable a seabed Treaty must provide *new* regulation for the ocean floor beyond the limits of national jurisdiction. This does not imply in any, however, that the new must be conceived in disjuncture from the old or in conflict with it: that the ocean floor must be regulated and administered in a vacuum. On the contrary, the structure of the ocean regime must provide a forum for dialogue between the old and the new. And this postulate has two dimensions, a territorial and a functional one. In territorial terms it means that while the old rights accruing to States from their sovereignty over a coastal zone must be respected, the new regime must provide procedures to solve, by common accord, problems circumscribed by ecological boundaries which do not coincide with political boundaries. In functional terms it means that while traditional uses of the sea, such as fishing and navigation, will continue to be regulated by traditional law, the new regime must provide dialogue between old and new users of ocean space, to reconcile conflicting uses and facilitate harmonious, balanced planning.

A seabed Treaty must cope with the exigencies of the scientific and technological revolution. The unprecedented role which science and industry play in the exploration and use of the ocean environment must somehow be reflected in the structure of the regime. New ways must be provided for their participation in decision-making and planning. This may well imply a new status under international law for nongovernmental or intergovernmental entities with new responsibilities and, very likely, a new standing before a maritime court or ocean floor tribunal.

A seabed Treaty must be an effective instrument of disarmament and the internationalization of controls: not colluding with illimited destruction by the perfunctory barring of limited destruction, but through the evolution of new concepts altogether, based on the recognition that the industrial uses and the military uses of the oceans are conflicting. By pushing peaceful, industrial uses of the oceans it may be possible to advance a system of peace in which the arms race simply would have no place.

A seabed Treaty must come to terms with the revolution of international relations. It must reconcile the requirements of the developed nations for stability and conservation with the demands of the developing nations for equal participation and sharing in the common heritage of mankind.

Only if it performs these four functions will a seabed Treaty be politically acceptable and, if accepted, workable.

The alternatives are gloomy. Undoubtedly, pollution would go unchecked. The forecast of the scramble for the wealth of the oceans leading to traditional imperialistic wars with untraditional, apocalyptic means may be unrealistic, as any projection of the past into the future. It might well be that the days of this sort of war are over, even though the military have not got the message yet. But the new kind of warfare, the guerrilla world civil war of today, transferred and perfected under the oceans, would be at least as terrifying. Because the more complex and sophisticated a system is, the more vulnerable it is. Imagine the guerrillas of the Third World, cheated out of their fair share of the common heritage of mankind, hitting at the nerve centers of the marine ecosystem. A crack in a half-million ton oil tanker: who will prevent it? Or the high-jacking of atomic weapons. Or cutting the lines of energy and oxygen supplies to submarine habitats and enterprises.

As the marine revolution proceeds, the lack of a legal regime, based on the political consensus of the world community, becomes more and more intolerable.

A successful seabed Treaty, however, provides for a constitution for the oceans that points in the direction of a constitution for the world.

There are no precedents. Reliance on existing models must therefore be limited. Innovation is called for.

Our choice is not between a "moderate" *status-quo* oriented regime, and a radical, utopian one. The *status quo* is the most unreal of all unrealities. Those who timidly aim at a "moderate" regime simply will not be able to sway the forces of inertia. There will either be no regime at all or there will be a comprehensive one—comprehensive in every sense of the word, and based on the necessary political and intellectual courage and passion.

These pages are indebted to the hundreds of participants in the Pacem in Maribus project in all parts of the world, whose contributions, written and oral, they freely incorporate, and whose thought is more adequately represented in the following chapters.

ELISABETH MANN BORGESE

PART ONE:

THE ECOLOGY

OF THE SEA

The law of the oceans must be in accord with the ecology of the oceans. Otherwise the law will perish—or the oceans. The instrument to join ecology and law is science policy.

The material gathered on this subject by the Pacem in Maribus project runs well over six hundred pages. This brief presentation cannot represent it adequately nor can it do justice to the immensity and complexity of the subject.

After a general introduction to the field as a whole (Carleton Ray), these pages concentrate on a set of particular problems and case studies.

The particular problem we have chosen is that of the "boundaries" of the marine eco-system: the interface between the marine eco-system and other global eco-systems and the legal consequences of these interrelationships for an ocean regime.

Norton Ginsburg explores the interface between marine and terrestrial systems, i.e., the coastal zones, where the confluence of ecologies produces maximum potential and maximal danger. This portion of the common heritage of mankind, this source of pollution for the entire marine eco-system, is under national jurisdiction. Whether it extends over three miles from shore or over two-hundred—and in ecological terms the first three miles are far more important than the remaining 197—it is clear that what happens there may affect the world community as much as the coastal na-

tion. Nations must cooperate to protect the common heritage against this source of corruption. Cooperation—while not abridging national sovereignty over the area in question—could take several forms. Codes and conventions could be proposed for ratification to all member states of the regime, regulating safety and pollution standards anywhere, within *the limits of national jurisdiction—just as conventions and codes have been adopted on human rights, slavery, labor standards, etc. Not as though the human rights convention should be considered as a shining example. If pollution standards were to be breached the way human rights have been trodden down over the past decades, the oceans would be dead in a few years time.*

Ratification of such a code or convention should be a condition for membership in the regime. The regime's deliberative body should review each year the activities of nations with regard to the code or convention (as the International Labor Office does with its conventions), and violators should be excluded from the regime: a sanction that should become, with the passing of time, as unbearable as exclusion from the Postal Union. The deliberate body of the regime should be entrusted with the responsibility of reviewing and revising the code or convention periodically, to bring it up to date in accordance with technological developments and requirements. While not infringing the sovereignty of any nation, the deliberative body should be able to make "recommendations" and render "opinions" addressed to nations with regard to their management of littoral resources and environment.

Wendell and Brooke Mordy deal with another ecological "boundary": that between water and atmosphere. They come to very similar conclusions with regard to the functions of the ocean regime.

The objective of the regime for the development, conservation, and security of the high seas and the seabed can only be met if the interdependency of ocean and atmosphere is fully taken into account, they state. They recommend that an ocean regime should have the explicit responsibility of rendering recommendations and opinions affecting the total global atmosphere. This is all the more urgent as there exists today no adequate body of municipal law to deal with problems which arise as we begin to use the atmosphere as a manageable resource.

We are rapidly moving from a phase of weather observation and prediction to one of weather control and modification—and the problems involved in this are far more political *than* technological, *as our experts point out. Ocean-based buoy systems are an essential part of this process. The law governing the ocean base affects developments in the atmosphere. This must be reflected in the structure and functions of the ocean regime.*

The case study we have chosen is the pollution of the Baltic, in which natural evolution and technological evolution seem to connive. The fate of the Black Sea, the Baltic, and the Caspian is in store for the Mediterranean and other parts of the marine ecosystem. Unless we act before it is too late.

The last two Chapters of this Part deal with science policy: past experience in international scientific cooperation, the existing scientific "infrastructure" for an ocean regime, and the direction in which it could be improved. We have chosen one American and one European expert: both agree on the need for institutional innovation and new forms of dialogue between industry, science, and government.

CARLETON RAY

1 Ecology, Law and the Marine Revolution

Mankind cannot attain a worldwide high standard of living without vastly increasing the use of the earth including the sea. Ecological facts of life are beginning to be accepted, but man has not yet solved the age-old paradox upon which his civilizations have many times foundered; namely, high population numbers with high cultural levels demand high environmental productivity, yet exploitation of nature produces environmental destruction and ecological collapse. When numbers of humans will exceed carrying capacity everywhere on earth, as is already the case in many nations, no one can say; but if man does not learn lessons of past history, there is no doubt that it will occur soon. The survival of man depends upon how he handles this challenge.

Five problems dominate the marine part of this challenge: (1) the development of international law with enforcement for exploitation of the sea, (2) the development of eco-system-based exploitation and conservation practices, (3) the cessation of existing destructive practices, (4) the assessment of marine environments relative to the carrying capacity of earth for man, and (5) the creation of marine parks, sanctuaries, and control areas for research. These are ecological problems which to date have been attacked in a piecemeal fashion. Ultimately, the answers will depend upon

---

CARLETON RAY is an Associate Professor, Department of Pathobiology, The Johns Hopkins University, in Baltimore. The full text of this paper was first published in *Biological Conservation*, Vol. III, No. 1, pp. 7-17, October 1970.

value judgments about what sort of world we wish to live in. Osborn (1953) asks: "Is the purpose of our civilization really to see how much the earth and human spirit can sustain?"

This chapter considers biology and law as they reflect upon the marine revolution. Biology and law require different approaches. The body of law by which we exercise control and responsibility is of man's creation. It should reflect common sense and be capable of rational alteration. Natural phenomena may make no "sense" at all and their complexities are infinite. It has been stated that the eco-system is not only more complex than we think it is; it is also more complex than we can think. The ecologist rarely can be definitive. He often experiences great difficulty in expressing, even to physically and experimentally-oriented scientists, the nature of the ecological crisis. Paul Erlich's (1969) "Eco-catastrophe" sounds to many like alarmist stuff.

To a great extent we are slaves to our history. The laissez-faire spirit of exploitation, the goal of economic growth, the socio-religious belief in man versus nature, and the conflict and case-history method of law make little sense when applied to the environment. The emerging marine revolution poses challenges to those concepts which magnify the importance of the sea far beyond its resource value. The wide recognition that this is so is reflected by the numbers of recent symposia and reports on the exploration, use, and legal regimes of the sea. Unfortunately, meetings of the American Bar Association and the Marine Technology Society have been composed almost entirely of industry representatives, lawyers, and a scattering of government officials, naval personnel and fisheries biologists. The latter mostly represent mission-oriented governmental agencies or industry. Marine ecologists are virtually absent!

Despite this the intensifying debate has produced the germs of workable ideas. The eco-system approach is just over the horizon and the greatest present need is for marine ecologists to make their voices heard. If they add their expertise to the debate, it is possible that nondestructive and cooperative exploitation on an international basis will replace the provincial madnesses acquired by land and perhaps marine eco-systems will not suffer more than they already have.

THE MARINE REVOLUTION

Man's massive entry into "inner space" initiates a marine revolution. It is producing not only resources, but new regimes for law, politics, and socio-economics as man investigates, uses, and hopefully will conserve that greater part of earth which has been mostly foreign to him.

Agricultural and industrial revolutions

Some thousands of years ago, man began to grow his own food. This change from the hunter-gatherer to the agriculturalist is termed the "agricultural revolution." It led to the urbanization and diversity of occupations which mark human culture. The agricultural revolution produced more food in a more accessible form than was available to the hunter-gatherer. Food, which presumably had been a limiting factor, was limiting no more. The carrying capacity of land for humans rose and the population grew accordingly.

The industrial revolution has been going on for the last two centuries or more. It has been marked by the growth of science and technology, by increased resource use, and by expanded diversity and efficiency of human skills. It has meant a turning away from the agricultural way of life to an increasingly urbanized and, perhaps we could say, the "artificial" one. It has once again increased the carrying capacity of the land for human beings and has lead to a spectacular increase in death control without concomitant birth control. Most significantly of all, the Industrial Revolution, in its greed for resources, has produced environmental destruction at an astounding and dangerous pace. Forests have been cut, land has been eroded and stripped, bays have been polluted and filled, and the result of all of these and other activities has been to lower the carrying capacity of land for future human populations. Such environmental wastage makes sheer hypocrisy our wish to provide a better life for our children.

The marine revolution

Thus does man turn to the sea which increasingly becomes vital for resources. However, the marine revolution is not totally a consequence of the exhaustion of the land. Man also turns to sea as it lies before him in the form of a challenge which he is now becoming technologically able to accept. "Products are sold on an open world market that cares nothing about the origin of the material; one competes only against price." (Bascom, 1966).

Thus, we accept the challenge of the sea, not a little starry-eyed over our technology. But we must remind ourselves that man remains the hunter-gatherer by sea; in only an insignificant few places does he farm. This contrast between advanced technology and the inadequacies of cultural and legal frameworks for regulation is a characteristic of *revolution.*

The marine revolution is quite as important a development as the previous agricultural and industrial revolutions. It is no more obvious on a day-to-day basis than the agricultural and industrial revolutions were in their time. Future man will clearly see this revolution as his inner-space logistics and utilization increase.

ECO-SYSTEMS AND HOMEOSTASIS

The eco-system is the fundamental functional unit of the natural world. It is comprised of all the living and non-living components of an environment and the totality of their interrelationships. An eco-system has properties of self-sustainment. Solar energy must be added, but nutrients and other materials are recycled. Examples are a lake, a forest, an estuary and a coral reef.

Carrying capacity, limiting factors and synergisms

Carrying capacity may be defined as the number of individuals of a species within a particular eco-system beyond which no major increase in numbers may occur. It fluctuates about an equilibrium level and may change seasonally or even daily. It is regulated according to Liebig's "law" of the minimum and Shelford's "law" of tolerance which together state that the presence or abundance of an organism is determined by the amounts of critical materials or by the levels of environmental factors such as salinity or temperature.

Typical of ecology, "laws" are easy to state, difficult to prove. A major reason for this is synergism; that is, environmental factors often act together to produce effects which are different quantitatively or qualitatively from the effects expected separately or additively. Carrying capacity and limiting factors apply to all living things. The foolish assumption is that technology may negate them for man. Technology cannot alter natural law; it can only redirect utilization in limited ways.

Productivity

Productivity is determined by turnover rate. The standing crop or biomass is a poor indicator as it tells little about how often materials are recycled. Plants absorb about one percent of solar energy for photosynthesis. An examination of trophic levels from these producers to primary, secondary, or tertiary consumers, reveals that each step involves about a ninety percent loss of energy. Thus, food chains are usually short and each trophic level is composed of a much smaller biomass than its predecessor.

Nutrients, unlike energy, are recycled. The bio-geochemical cycles of nitrogen, phosphorus, and minerals are most efficient in complex eco-systems. Man can occasionally increase productivity through the addition of substances which once were limiting. More often, his "making the desert bloom" fails in the long run through failure to recognize the interrelationships of these cycles.

Primary productivity varies widely. Deserts and the waters of the deep oceans, which together cover most of the earth, produce less than one gram of dry organic matter per square meter per day. Grasslands, waters over the continental shelf, and marginal agriculture produce half a gram to three grams; moist forests and agriculture produce three to ten grams; estuaries, inshore seas, and intensive agriculture produce ten to twenty-five grams. (Odum, 1959.)

The seas contain more living material than the land because of their large total productive area and volume. However, man's use is at a higher trophic level in the sea; land = sun → grass → cow; sea = sun → phytoplankton → zooplankton → primary carnivore (herring) → secondary carnivore (tunny). The sea contains a much greater diversity of life than the land, but because oxygen content of water is lower than that of air, the sea is dominated by animals of low metabolic rate. Last, the sea is more a stable environment than the land; "weather" is mild and the productive season is long. For all these reasons, productivity by sea is not as it is by land in kind or amount.

Homeostasis, simplification, and pollution

Homeostasis defines the "balance of nature." All eco-systems depend upon recycling for sustainment and upon complexity for stability. These involve intricate mechanisms analogous to (but more complex than) the heat producing, dissipating, and conserving mechanisms which regulate human core temperature. Eco-systems are never perfectly balanced, but homeostatic mechanisms give them recuperative power which, when exceeded, leads to breakdown; the eutrophication of Lake Erie is a classic example.

A major part of homeostasis lies in complexity which insures both productivity and stability (Dasmann, 1968; Elton, 1958). Man is a simplifier of eco-systems and thus reduces their recuperative power. The many forms of pollution are the most serious stresses. Historically, man has depended upon maximum homeostatic capacities of the environment to endure pollution, but in simplifying and polluting at the same time, he attacks with a two-edged sword.

Is the ocean too large to disrupt? The author thinks not. According to the Task Force on Environmental Health and Related Problems in 1967, the American people and their environment are being exposed to half-a-million different alien substances with twenty thousand new ones added each year. Some of these go to sea. For instance, pesticides have been distributed throughout the world's oceans through the vectors of air and precipitation (Frost, 1969). Polikarpov (1966) suggests that radionuclide pollution may already be at a maximal level in seawater and it would be difficult to prove otherwise. Hedgpeth (1970) remarks that our standards for waste disposal are anthropocentric and that laboratory tests on pollutants are "interesting, but possibly academic as far as the real world is concerned."

MAN'S USE OF THE SEA

Only recently has man begun to explore the sea throughout its three dimensions. The first extensive exploration of the deep sea was in 1873–1876 by HMS *Challenger*. Not quite a century later, man has visited the ocean's deepest place and knows that all marine waters are capable of supporting life.

The marine revolution consists of five major parts: fisheries, minerals and mining, military interests, science and technology, and conservation and recreation. Emery (1966) gives world values of marine resources in 1964, as follows: biological–$6.4 $\times 10^9$; geological–3.6 $\times 10^9$; and chemical–1.3 $\times 10^9$. Biological resources will always be the most valuable, even if surpassed economically, for man cannot exist without them.

Fisheries

Fisheries remain the most difficult aspect of international law of the sea. This is mainly because of the fact that most organisms move and cannot be claimed. It is ludicrous to discover that certain benthic organisms are, in fact, classified as "minerals" under the Convention of the Continental Shelf. In some cases, it is of advantage to the exploiter to do so, for instance, Alaska King Crab (Oda, 1968), and in some cases not, for instance, shrimp (Neblett, 1966). Fisheries resources include algae, plankton, shellfish, fishes, turtles, and mammals (Walford, 1958); but, as has been pointed out above, man's utilization represents only a fraction of total marine productivity.

Over-utilization continues to dominate fisheries, especially offshore ones. Clark (1967) states that Japanese long-lining accounted for almost a million billfishes in 1965. Even larger quantities of tunny were taken. Evidence is accumulating that such utilization cannot be sustained. Perhaps even more serious than overfishing is inshore habitat destruction. Over two-thirds of all commercial and sport fishes of the eastern United States depend upon inshore and environments at some critical time of their life cycle. The most effective way to extirpate a species is by environmental disruption, and this is being done inshore at a rapid pace.

Consideration of energetics leads many to propose exploitation at lower trophic levels. Complex size/metabolic factors and fishing efficiency strongly indicate, however, that higher order consumers are more effective fishermen and converters of energy than man is. A total "plankton" fishery should be considered a last, and none too satisfactory, resort. Those who have taste-tested swordfish and plankton might agree! The choice, however, should not be between swordfish and plankton; given proper management, we could have both.

The concept of "yield" is vital biologically and legally. Fisheries biologists have emphasized the asymptotic attainment of maximum biomass through controlled utilization. Such a yield may or may not conform to economic efficiency or to local market value, hence the preference of "optimum" over "maximum" yield (Crutchfield, 1968).

Chapman (1966) states an exploitative point of view: "When the fishing effort has increased beyond the point of maximum sustainable yield, the fishing can ordinarily be permitted to expand without serious damage to the resource." He ignores Allee's principle which is that density is in itself a limiting factor for population growth and survival. Relative abundance between species is a contributor to homeostasis. Thus, it is biologically most sound to change population size as little as possible in natural systems.

Christy (1966) considers broader aspects of utilization: ". . . somehow or other it will be necessary to limit the number of fishermen that can participate in a fishery. Such limitations can only be achieved by further restricting the 'freedom of the seas': and this clearly raises questions about the meaning of this freedom and about the distribution of wealth." This approach appears to me more susceptible to ecological application than Chapman's more narrowly stated views.

Aquaculture presents different sorts of problems than hunter-gathering and may be the dominant provider of the future. Aquaculture is a major concern of the U.S. Sea Grant Program (Abel,

1968). Rhyther and Bardach (1968) and Bardach and Rhyther (1968) review aquaculture and make the point that it will be carried out largely along coasts—exactly the areas currently most stressed at the hand of man. To reconstitute coastal environments, or to artificially fertilize them is difficult or impossible. The key to aquaculture is clearly the maintenance of natural productivity.

Minerals and mining

The literature in this field often leaves one impressed with the viewpoint that somehow we are slaves to "economic growth." Close (1968) speaks of "the care and feeding of a gigantic industrial complex." One hopes that only a segment of industry would speak so carelessly, but it does appear true that an awareness of ecology and a willingness to exploit the nonliving resources at little or no expense to the living are indeed rare. If mineral exploitation continues by sea as it has by land, the predictable results are frightening to contemplate.

Mero (1966, 1968), Luce (1968) and Young (1968) review the diversity of mineral resources in the sea. Inshore mineral exploitation is already heavy, but a consensus exists that only a few minerals are presently feasible of exploitation, exceptions being oil and gas. This is evidently based upon the lack of a favorable legal and economic climate, not upon the lack of technological capability. Further, it is not true that exploitation will progress from shallower to deeper water, this being a function of the resource sought (Wilkey, 1969).

Offshore mineral production in 1968 was six percent of the world total of which oil and gas were eighty-four percent (Economic Associates, 1968). In 1965, sixteen percent of the free world's oil was produced offshore, the result of the work of 325 rigs which have drilled many thousands of wells (Dozier, 1966); oil is being produced from wells in 104 meters of water (Wilkey, 1969) and exploratory drilling was carried out in 1968 in the Gulf of Mexico in over 3600 meters. At any one time, about thirty million tons of oil are at sea in tankers. From U.S. offshore wells alone the production of oil has so far been two billion barrels and of gas, 5.5-trillion feet, at an investment of six billion dollars, and with the ultimate potential of fifteen to thirty-five billion barrels of oil and ninety to 170-trillion feet of gas (Nelson and Burke, 1966). The massive pollution potential of the oil industry has been previewed by the tragic *Torrey Canyon* and Santa Barbara disasters. There is certainty that these episodes are not the last.

Military interests

Unfortunately, the military is shrouded in secrecy. It would, for instance, be interesting to find what the degree of radionuclide polution is from Soviet and U.S. nuclear-powered submarines. Both Harlow (1966) and Hearn (1968) state the Navy's conservative position, that maximum freedom to use all dimensions of the sea must be maintained in order to exploit naval strength to the fullest in the best national interest—and it is only fair to state that such a position is shared by the military of other major powers. The effect is to raise a serious obstacle to internationalization, to expanded territorial jurisdiction, and to peaceful use of the sea floor.

It is difficult to understand why putting the seabed under a "peaceful purposes only" treaty, as has already been done for outer space and Antarctica, is not in the "best national interest." Evidently military influence was a major factor in preventing that principle from being accepted at the 1967 UN debate on the subject (Eichelberger, 1968). And it is particularly disturbing to read that "military strategists . . . have been looking for better ways to put the sea to use for the purposes of national defense" (*New York Times*, 1969).

It must be pointed out that military interests are not necessarily contrary to fishing or mineral exploitation. By no means should international progress on the latter be held up as a result of conflicts with the military.

Science and technology

The United States is heavily committed to marine exploration, science, development, and conservation. Reports on the highest level are numerous, including: Interagency Committee on Oceanography (1963, 1967); National Academy of Sciences (1964, 1967, 1969); Panel on Oceanography, President's Science Advisory Committee (1966); National Council on Marine Resources and Engineering Development (1967, 1968a, 1968b); and the Commission on Marine Science, Engineering and Resources (1969).

The last mentioned, the so-called Stratton Commission Report, departs courageously from its predecessors and is no doubt the most significant. It is ecological and international in nature and recommends a U.S. National Oceanographic and Atmospheric Agency for centralization of U.S. research, exploration, data collection, and education.* Further, it proposes an International Registry Authority for ocean claims with regimes for ocean bottoms, a delineated continental shelf, and an intermediate zone. The Commission also stresses optimal use of coastlines on a long-term basis in which industry, water quality, and aquaculture would be regulated under federal law to guard against deterioration of the inshore marine environment (Clingan, 1969; Lawrence, 1969).

Looking not at reports, but budgets, produces some dismay. *Ocean Science News* (1969) states the current federal commitment to marine matters to be $528 million of which but $150.6 million is in basic and applied research and $143 million in national security—this in the same year of man's travel to the moon! The over-all oceanic budget has grown twenty-two percent since 1968 when Economic Associates (op. cit.) remarked: "what remains to be pointed out is the very low level of federal expenditure on . . . resources and their environment compared with federal oceanologic programs in general and, decidedly so, with the federal effort in such a field as outer space."

The International Biological Programme's Marine Productivity

---

* NOAA was established in 1971. It is composed of selected marine programs transferred from the Department of the Interior, the National Science Foundation, the Coast Guard, the Navy, and the Army Corps of Engineers. The budget totals about $270 million, and it has a personnel strength of over 12,000. The following programs are transferred to NOAA: Environmental Science Service Administration (Commerce); most elements of the Bureau of Commercial Fisheries; the marine sport fish program from the Bureau of Sport Fisheries and Wildlife; the major part of the anadromous fish program, and the marine mining program (Interior); the office of Sea Grant Programs (NSF); the national data buoy development program (Coast Guard); and the Great Lakes Survey (Corps of Engineers). NOAA is located in the Department of Commerce. *Ed.*

section deserves mention. The IBP theme of "The biological basis of productivity and human welfare" is ideally suited to the needs of man during the initial period of the marine revolution. However, at the current level of funding (less than five million dollars for all U.S. IBP sections in FY, 1970), it is certain that IBP cannot fulfill its goals.

Conservation and recreation

To many, conservation and recreation mean the establishment of parks, sanctuaries, and control areas for research (Chapman, 1968; and Ray, 1961, 1965, 1966, 1968). However, conservation and recreation must not be confined to protected areas. Both must principally be concerned with the maintenance of eco-system homeostasis on a worldwide basis and this is a large order indeed.

The concepts of conservation have been developed for terrestrial environments and are only vaguely applicable to the sea. The sea is a vastly larger biosphere than the land and is more continuous. Its rate of change, its biotic complexity and our ignorance of its three-dimensional hydrosphere are of a different order of magnitude than for the more familiar land. For both land and sea modern conservationists have become less concerned with the placing of fences about sea or landscape, valuable as protective measures are, than with an ecological concept of the total ecosystem with man as a part. A good base of conservation policy exists for land and, in part, for inshore seas. For the high seas, this is not the case.

LEGAL REGIME OF THE SEA

Ultimately, man's activities within and beneath the sea must be legally regulated. Griffin (1967) states: "To a large extent, a period of legal conjecture is ending." The problem is ". . . to evolve policies and a legal regime which will maximize all beneficial uses of ocean space. . . . Under no circumstances, we believe, must we ever allow the prospects of rich harvest and mineral wealth to create a new form of colonial competition among the maritime nations." A contrary view is that of Ely (1967a): "Above all, we should not now cede to any international agency whatsoever the power to veto American exploration of areas of the deep sea which are presently open to American initiative. We can give away later what we now keep, but the converse is sadly false." Ely (1967b, 1968) extends these views.

Basically, the argument concerns whether the sea and sea floor are *res nullius* (belonging to no one and subject to claim) or *res communis* (property of the world community). Put another way, Eichelberger (op. cit.) says: "Either [the sea] opens up another threat of conflict or another area of cooperation." Of course, the argument is not so simple. As Friedham (1966) and Belman (1968) point out, traditional law of the sea is imperfect, but there is legitimate hesitancy towards creating new modes when our experience with the sea and our ignorance of its resources is still great.

Historical background

In 1609, Grotius wrote "Mare Liberum" as a challenge to national jurisdiction of areas of ocean. This brief for the Dutch Government was directed towards breaking the Portuguese monopoly of the East Indies spice trade. Gradually and in partial response to struggles for supremacy between Britain and Spain, the principle of "freedom of the high seas" was accepted.

The concept of a territorial sea was born when Bijnkershoek wrote "De Dominio Maris" in 1702. A territorial width of three miles had been attributed to the distance of a cannonball shot, but the range of cannon at the time was only a mile. Probably the three-mile limit began with a British instruction to its ambassadors in 1672, that control should be exercised one marine league (equal to three nautical miles) from shore (Weber, 1966). Three miles was never adopted universally; claims of up to twelve miles have always been valid.

A Convention of 1884 sustained all states rights to lay cable on the deep sea floor, but it was not until the Treaty of Paria between Britain and Venezuela in 1942 and the Truman Proclamation of 1945 that any state claimed jurisdiction and control over any part of the sea floor. By its important action, the United States effectively laid claim to an area of shelf larger than Alaska and Texas combined.

Three and a half centuries of precedent thus led to recognition of the following zones: (1) internal waters and bays within the control of the coastal state; (2) territorial sea under the control of the coastal state; (3) continental shelf over which the coastal state might claim control; (4) contiguous zones for special purposes; (5) the high seas, held to be *res communis*; and (6) the deep sea floor, held to be *res nullius.* New technology for ocean research and exploitation after World War II indicated obvious conflict under this system.

The International Law Commission had been created in 1947 under the United Nations. It proposed in 1956 that a Conference on Law of the Sea be held. This occurred in 1958 at Geneva. It adopted four conventions (cf. texts: American Bar Association, 1967):

1. Territorial Sea and the Contiguous Zone. Came into force September 10, 1964. This convention confirmed the control of the coastal state over all resources within a territorial sea. In addition, the coastal state might declare control over a contiguous zone for security, customs, fiscal, immigration, or sanitary purposes but not interfere with right of innocent passage. The width of the territorial sea is still undecided. Of ninety-one coastal states, fifty declare twelve miles, seventeen declare more than twelve, ten declare between three and twelve, and fourteen declare three miles (Oda, op. cit.). A narrow territorial sea is favored by the military and advanced fishing states; Japan is the only major fishing nation which adheres to three miles. A wide territorial sea is favored by states wishing to protect a coastal fishery. Obviously, the U.S. has been in a delicate position and declared twelve miles only recently.

2. High Seas. Came into force September 30, 1962. This includes all waters outside territorial ones and declares freedoms of navigation, overflight, fishing, and the laying of submarine cable and pipeline. Also included are regulations on piracy and pollution.

3. Continental Shelf. Came into force June 10, 1964. This convention is mainly concerned with the sea floor and does not include

the water above. It has already been pointed out that certain living resources are included. The most serious contention concerns the extent of the shelf: ". . . to the sea bed and subsoil of the submarine area adjacent to the coast, but outside the area of the territorial sea, to a depth of two hundred meters or, beyond that limit, to where the depth of the superjacent waters admit of the exploitation of the natural resources of the said area." Two schools of thought prevail here. One contends that because this convention is entitled "Continental Shelf," the sea bottom beyond its geographic limits of about two hundred meters depth is not included. The other contends that the exploitability provision defines a "juridical shelf" which could include the slope or even the whole ocean bottom. It should be kept in mind that the shelf area is a huge one; without the slope it comprises ten million square kilometers,[2] equal to twenty percent of the total land area of earth (Mero, op. cit.). An excellent review of the problem is that of Tubman (1966).

4. Fishing and Conservation of Living Resources of the High Sea. Came into force March 20, 1966. This remains the most controversial of the conventions, the only one which did not more or less standardize a body of existing custom but which contained genuine innovation. The problem that one noncooperating state could vitiate fishery conservation efforts was a major reason for calling the Geneva Conference. This convention "virtually forces consideration of the need for conservation of a fish stock by all participating nations if only one (or an adjacent coastal state) insists on it," but "it says nothing about the principles to be followed, nor, more fundamentally about the objectives sought" (Crutchfield, op. cit.). It does not treat allocations nor provide more than case-by-case consideration of conservation.

Prognosis

Christy (1968) outlines four approaches to the developing law of the sea. The "wait and see" approach leaves exploitation to chance. Support for wait-and-see comes in part from proponents of case law who heed the dictum of Oliver Wendell Holmes: "The life of the law is not logic, but experience." Additional support accrues from those who note our lack of knowledge and experience in the sea.

The second approach is that of the "national lake." The obstacle here is that the division of the sea would be highly inequitable. The USSR would get little, whereas tiny oceanic islands would gain title to huge territories.

The "flag" approach is the third. It is supported mainly by mineral and military interests of powerful nations. McDougal (1968), Wilkey (op. cit.) and Burke (1966a, 1966b, 1968, 1969) defend this point of view, emphasizing traditional processes of mineral claim on and under a seabed held to be *res nullius.* Some are willing to make concessions on an international registry or towards cooperation in pollution and security. Young (op. cit.), Krueger (1968), and Eichelberger (op. cit.) hasten to point out that the flag approach is but a form of neo-colonialism which would rapidly lead to a gold-rush. Nor does the flag approach, with its unavoidable competitive nature, make much sense ecologically.

The last alternative is "international." Krueger (op. cit.) and Eichelberger (op. cit.) lucidly point out the obsolescence of nationalism and the fact that the multitude of small nations will view internationalism as the only legitimate approach to the sea. Further, mineral resources required by the industrial nations are spread throughout the international market necessitating international trade.

The United Nations has shown its resolve by a series of resolutions. One of December 31, 1968, designated Resolution 2467A-2467D (XXIII), includes the following points: (1) promotion of international cooperation; (2) exploitation for the benefit of mankind; (3) prevention of pollution; (4) desirability of peaceful use of the seabed; and (5) endorsement of an International Decade of Ocean Exploration. The author finds it impossible to argue against any of these goals and equally impossible to see an alternative to internationalism in achieving any of them. Precedents of treaties on Antarctica and outer space exist though both Young (op. cit.) and Eichelberger (op. cit.) point out that the ocean floor is not *tabula rasa* as were Antarctica and outer space. However, they do not point out that virtually all of Antarctica was under territorial claim and that nuclear testing and exploration had been carried out in outer space when those treaties were signed. Both treaties involved a yielding of claims and nullifications of military interests. It is difficult to see why such yielding could not also take place for the sea floor, the superjacent waters, and even some sections of shelf. One thing is certain; under no reasonable circumstances would the exploiter lose by international control. All that might ensue would be more efficient utilization and a cleaner sea.

Gargantuan problems exist with regard to internationalism. Burke (1966a, 1966b, 1968, 1969), Alexander (1966) and Griffin (1967) review the problems of disarmament, bilateral and multilateral agreements, the extent of offshore claims, scientific freedom in research, and many others. Burke (1969), particularly, examines difficulties in applying the Stratton Commission Report. However, one cannot be deterred from a path simply because it is stony.

SUMMARY

The attainment of a worldwide high standard of living depends upon vastly increased resource exploitation, including the sea. Man's exploitative activities heretofore have simplified environments, reducing their homeostasis and leading to eco-system collapse. Eco-system ecology is exceedingly complex and the ecologist can rarely be definitive. However, the system of law by which man regulates his activities is of our own creation and should be subject to change according to need.

Many historically developed modes of human activity make little sense ecologically. The marine revolution follows the agricultural and industrial revolutions as a significant change in man's relationship with his environment. Advanced technology for ocean research and development are becoming available, but cultural and legal frameworks for regulation have not matured. A major problem in the development of a marine tradition whereby man will not destroy marine eco-systems lies in the application of ecological "laws" to our activities at sea.

The uses of the sea are fisheries, minerals and mining, military interests, science and technology, and conservation and recreation. Legal regimes for regulation stem mainly from the four Geneva conventions which mostly formalize a three and a half century history of marine law. These conventions themselves emphasize the recognition that marine law needs modification along new lines. Debate intensifies over the *res nullius* versus *res communis* regimes for ocean exploitation.

Exploitative "conquest" can no longer serve as a guide for man's use of the sea as it has for the land. An emphasis must be given to ecology, to eco-systems, and to the role of the marine ecologist in the oceanological debate. Concentrated thought is necessary along positive lines, whereas in the past provincialism and tradition have stood negativistically in the way of international control of ocean resource use. Should an over-riding consideration be given to ecology and internationalism, the marine revolution will affect man's history far more than an evaluation of resources alone would indicate.

CONCLUSION

The sea lies today like a huge plum which man is ready to pluck but towards which he gropes in quandary. This paper emphasizes the application of ecology to this marine revolution. We see that historically we have grown to treat the sea as the land—with exploitation and as a "frontier" to be conquered. That this is a collision course and that the "conquest" of nature threatens man's existence as a species with high "culture" is no longer in doubt.

Much as we might wish it so, the sea is not a placebo for our destruction of the land. The very existence of conventions on the sea are cause for optimism and proof of awareness of the need for change. To the international lawyers belongs most of the credit. However, there persists the house lawyer's fear of loss of proprietary rights, the industrialist's fear of loss of claim, the fisherman's fear of loss of laissez-faire exploitation. Many maintain that we do not yet know about the sea nor do we have enough experience with it to change our *modus operandi*. Nevertheless, one must agree with Belman (op. cit.): "If law awaits developments, it loses the ability to shape them."

The eco-system principle must serve as the overriding guide for shaping our future resource use. We simply do not dare approach limits of homeostasis in the sea. Ripley (1966) states: "The basic problem therefore is to acquire sufficient knowledge about our eco-systems to provide feedback controls essential to homeostasis." It is true that we do not as yet have all the knowledge we might desire, but it is also true that we know enough now to be able intelligently to monitor our actions. We *can* assume that every one of our actions puts some stress on the environment. We *can* put aside expediency, tradition, and false economic idols. We *can* negate flimsy and obsolescent national boundaries. We *can* shift the burden of proof for ecological damage from the plaintiff-community to the defendant-exploiter. The problem is not ability to change; it is desire.

A new brand of environmental biologist must become increasingly involved in the marine revolution; without him, no purely political or legal solution will suffice. Non-biologists, even lay conservationists, have too rarely shown comprehension of the complexities of the living world and they cannot deal alone with the sophistication of eco-system ecology. However, the biologists have been too rarely willing to commit themselves. Darling (1967) has pinpointed part of the problem: ". . . public policy has to be ahead of public consensus . . . ecology and conservation can move surely into the hurly-burly without losing scholarly integrity, a course most of us must be prepared to follow. . . ." Biology must, to a new degree, achieve interaction with politics and the law. Scientific integrity must be defended, but this is not in conflict with a willingness to become involved with the issues.

There is apparently no end in sight to man's reproductive potential nor to his infinite conceit that he shall inherit the (still productive?) earth. There is a limit to the sea as to the land. When will the curves of population and productivity cross? To put off the date with stop-gaps is no final answer. Palliatives and indecision will not suffice.

BENGT LUNDHOLM

2 The Oceans—Their Production and Pollution. The Baltic as a Case Study

The Baltic is the largest body of brackish water on earth. Its volume is 22,000 ^{3} kilometers, and its area is 370,000 ^{2} kilometers—compared with the 82,000 ^{2} kilometers of Lake Superior, or the 26,000 ^{2} kilometers of Lake Erie. The average depth is only sixty meters and the maximum depth is 459 meters.

There are several shore owners: Finland, the Soviet Union (USSR) with several republics, Poland, Eastern Germany (DDR), West Germany (BDR), Denmark, and Sweden.

The water balance is rather complicated and known only in general. The Baltic Sea is connected with the North Sea through Kattegatt and three entrances between Sweden and the Danish Islands: Öresund, The Great Belt, and The Little Belt. From the Baltic there is a surface current with a low salinity, but—and this is important—in the Belts there is a deep current with higher salinity and heavier water going in the opposite direction. Through its rivers the Baltic receives 471 ^{3} kilometers per year, and the same volume comes in through the deep current from the Belts. The surface current will contain the double volume or 942 ^{3} kilometers. As a result, there are two rather separated layers of water in the Baltic. The difference in salinity obviously is the cause of this.

The incoming deep current has a salinity of 17.5 o/oo (promille).

Bengt Lundholm is the Director, Ecological Research Committee, Swedish Science Research Council, in Stockholm.

The deep layer—11 o/oo (promille).
The surface layer—7 o/oo (promille).

Primary production in the water depends on several factors. The amount of nutrients is of importance and especially nutrients which occur in such a small amount that they are growth-limiting factors. Phosphorus is one such limiting nutrient. The amount of phosphorus in the Baltic surface water is between seven to ten micrograms per liter, compared with twenty-five to thirty-five micrograms in the North Sea. As a result, the productivity of the Baltic is low compared to that of the North Sea. In the bottom layers with high salinity, however, a very large amount of phosphorus is stored. This phosphorus pool has been estimated to be 325,000 tons. If we get a complete mixing of the two layers, the phosphorus amount should increase from eight to twenty-four liters. The surface water would then be rich in phosphorus and able to support an increased organic production. Only ten years ago technicians were claiming that if we could pour enough nutrients into the Baltic, the fish catches would increase, and that sewage from cities and industries would be a blessing. But this turned out to be wishful thinking on the part of the municipalities.

Consideration of the fact that organic production also depends on oxygen is important. The higher organisms in particular can not exist without oxygen. The term biological oxygen demand (BOD) is well-known in connection with water waste. When organic waste material passes through a sewage plant it is partly broken down. The nutrients or unorganic building blocks are freed. A complete breakdown of the organic material occurs in the recipient, and oxygen is necessary for this. So that is the primary oxygen demand.

Nutrients in the water are the base for new organic material which is created by assimilation of the green algae in the water. This results in water bloom. When the algae die, they are broken down by microorganisms which again need oxygen. This is the secondary oxygen demand.

Let us return to the situation in the Baltic and apply this knowledge. There we had two layers of water with a differing salinity.

The boundary between the two layers is called the halocline. Oxygen from the air penetrates down to the halocline. Above the halocline there is a seasonal inversion of the water resulting in an even distribution of the oxygen. However, it is very difficult for the oxygen to pass the halocline. If nutrients now get into the surface water, there is water bloom, and when the organic material dies it sinks down through the halocline to the deeper layers where it is broken down. In this case the oxygen is taken from the deep salty layer where the oxygen level was already low. This may result in a lack of oxygen in bottom layers of the Baltic. Then, if partly broken-down sewage is poured out above the halocline, the following results: the primary oxygen demand lowers the oxygen level slightly in the surface water, and the secondary oxygen demand has a more marked effect on the bottom layers. If the effluent is below the halocline, there might be a catastrophic decrease in oxygen. In the latter case, however, no one will find out—with the exception perhaps of some inquisitive scientist.

The very salty bottom water in the deepest parts of the Baltic is renewed in an extremely irregular manner. There is a large inflow of salt water after which the bottom water may be stagnant for several years. In this stagnant water the oxygen levels are decreasing partly because of inflow of organic material which then is broken down and uses the oxygen. Oxygen levels in the deeper part of the Baltic have been followed for the last seventy years. Even if there are changes resulting from new saltwater inflows, the trend to decreased levels is still very marked. There are dramatic changes in the water from oxygen to hydrogen sulfide, that is, a change from aerobic to anaerobic conditions. The hydrogen sulfide acts as a poison on all the higher organisms, and the bottom animals—so important as fishfood—will die and disappear. The fish avoid such areas, and earlier spawning areas are now deserted. The most important change, however, is that the reducing bottom sediments now will release large amounts of phosphorus. Thus the water will suddenly receive an increased input of nutrients from the sediments.

There was a marked increase in hydrogen sulfide on the bottom areas of the Baltic between 1966 and 1969. This increase resulted from a depletion of oxygen, probably caused by a transport of organic material to the deeper layers. This trend was broken however by a marked inflow of oxygen-containing heavy water with high salinity from the Belts. The Belt water pushed away water from the bottom hollows thus transferring the nutrients to the upper oxygen-rich layers.

Early in 1970 there was a rapid decrease in hydrogen sulfide on the bottom layers, and once again there is oxygen. However, there is still the important probability that oxygen levels will decrease more rapidly this time because large amounts of nutrients have been flushed away from the deep basins. We now can expect eutrophication of the entire Baltic. The immediate result will be more phytoplankton and an increased number of fish, but it is feared that the end result will be increased areas with hydrogen sulfide—areas which will poison the fish and ruin fisheries. The question is whether this situation is caused by a natural development with the end result to be the same as in the Black Sea where a layer of hydrogen sulfide dominates, or if it is caused by human activities and a possibility may exist for breaking the present trend.

A rough balance sheet of the phosphorus in the Baltic will show the magnitude of human influence. The yearly increase of phosphorus in the Baltic is:

Bottom layers	13,000 tons
Surface layers	2,000 tons
Increased biomass	4,000 tons
Total:	19,000 tons

From Kattegatt	3,500 tons
From precipitation	3,000 tons
From rivers	7,000 tons
Total:	13,500 tons

We have thus a balance of 5,500 tons which probably are from human activities—for the most part from municipalities and industries that use the Baltic as a direct recipient.

If we divide this input between natural and human sources, we get the following results:

Natural sources:	
From Kattegatt	3,500 tons
From precipitation	2,000 tons
From rivers	3,500 tons
Total:	9,000 tons

Human sources:	
From precipitation	1,000 tons
From rivers (detergents)	3,500 tons
Unidentified source	5,500 tons
Total:	10,000 tons

Therefore, the human and natural influences are of about equal size, but the important thing is that human influences have increased over the last twenty years and are still increasing. In addition, calculations show that the amount of organic material from the surface layers to the bottom layers is at present of such a size that the amount of oxygen in the bottom layers is barely sufficient to break down the organic material. An increase of organic material probably will result in permanent anaerobic conditions. From these calculations it is clear that an increase of human phosphorus will be disastrous.

Correlated with these changes caused by human activities are changes which might, at least for the present time, be regarded as natural. During the last seventy years the salinity in the Baltic has increased 1 0/00 promille. The reason for this is a decreased inflow of fresh water through rivers, probably caused by climatic changes. This has resulted in increased inflow of salt water along the bottom of the Belts. The halocline was situated at an average depth of one hundred meters around 1900. The present depth in the central Baltic is sixty meters. This increase of saltwater volume also has another effect. During certain weather conditions the halocline is tipped over and reaches almost up to the surface in coastal areas. On these occasions there is an increase in nutrients, which results in local eutrophication of coastal areas as well as an increase in the amount of vegetation and water bloom where no local pollution sources are recorded—all of which bring complaints from people using these waters for recreational purposes.

It is rather interesting that a remarkable transport of vegetation material, mostly different types of higher algae, have been found to occur from the coastal areas along the bottom to the deeper parts.

After these hydrological and chemical aspects, a few words should be said about the biology of brackish water. In the Baltic the number of species decreases with the salinity. The eco-systems in the Baltic are very young, and the species there are confronted with difficult problems of adaptation. They now live under conditions of constant stress, so their food webs are rather simple. The effects of additional stress will lead to further simplification. However, under conditions of pollution we could expect less stable conditions, but that remains to be proved.

The effects of particular pollutants also are of importance. There are areas where fish contain more than one part per million of mercury. These areas have been blacklisted in Sweden and fishing is forbidden. Also certain coastal areas in the Baltic have been blacklisted. This is of direct economic importance.

The levels of DDT and PCB are much higher in the Baltic than in the southern Baltic and the Kattegatt. Very high levels of these chemicals in eagles have resulted in infertility. The levels in fish are so high that if we applied the tolerance levels used for vegetables we would be in real danger of having to forbid fishing.

SUMMARY

The Baltic is the largest volume of brackish water on earth. A surface current with low salinity flows from the Baltic. A deep current with high salinity flows to the Baltic as does the fresh water from various rivers. As a result there are two rather separated layers of water, a surface layer with a salinity of 7 0/00 and a deep layer with a salinity of 11 0/00. They are separated by a halocline, which prevents oxygen penetration.

The amount of phosphorus in the surface water is low, but the bottom layers contain a large amount of phosphorus. Productivity of the surface water is rather low. Organic material is continuously transported to the deeper layers where it is broken down with oxygen taken from the deep layer. The result of this may be a complete depletion of oxygen in the bottom layers, where instead hydrogen sulphide will appear. Over the past seventy years oxygen in the deep layers has decreased despite an extensive irregular ventilation of the deep water by inflow of oxygen-rich deep water. The result will be an increased organic production in the surface water. When organic material from this increase in production reaches the deep water, a rapid depletion of oxygen will occur.

A risk now exists that the Baltic has reached a point where, like the Black Sea, it will have permanent hydrogen sulphide in the deeper parts. From both the biological and practical points of view, this change may be of utmost importance to all nations adjoining the Baltic. In part, the change may have been caused by human influences such as sewage and waste materials.

We cannot use the seas both as cesspools and as productive areas. We must make a decision, and the decision must be international in scope.

NORTON GINSBURG

3 The Lure of Tidewater: The Problem of the Interface Between Land and Sea

The mark of man on earth's land surface is as old as his occupancy of the earth; but his mark on the seas is the product only of very recent periods in history. At the beginning of the nineteenth century, it was reasonable, if romantic, for Byron to write:

> Man marks the earth with ruin—
> his control stops with the shore.[1]

It may still be true that man continues to leave his ruinous mark on the earth itself, but neither his interest nor his control stops at the shore. Not only are men, more than ever before, concerned with the uses and resources of the seas but their capabilities to explore and exploit these resources are also greater. Although it is customary to think of outer space as the great new frontier for exploration, in fact that claim might even better be made for the oceans, the "unknown seas." On a world scale, then, questions concerning the nature of the oceanic eco-system, the rights of men and nations to explore and exploit it, must occupy our attention. Such questions justify the need for conferences on the oceans, and for the peaceful uses of oceans and their resources.

On a lesser scale, however, there are other problems which call for attention. In the past century, but most markedly in the past few decades, a distinctive geographical zone of occupancy and use

---

NORTON GINSBURG is a professor in the Department of Geography, University of Chicago and Dean of the Academic Progress Center for the Study of Democratic Institutions.

has come into being in various key regions of the world, a zonal interface between land and sea but including parts of both, forming a new kind of geographical, areal subsystem fraught with problems and possibilities. Less a world phenomenon than a national and regional one, the problems of this zone nonetheless have global implications. Our purpose here is to identify this subecosystem zone of interaction and relationship and to explore some of the problems associated with it.

First, let us imagine a map showing the distribution of population in the world. To a striking degree, most of the world's population lives within two hundred miles of the sea. Most of the continental interiors are relatively lightly populated, except where arms of the sea in the form of navigable rivers like the Yang-tze and the Mississippi plunge deep into the hearts of continents. Even so, such waterways do not guarantee high interior continental population densities, as the Amazon so well demonstrates. It is not strange that the population of the world should be distributed in this fashion. Most of the great fluvial lowlands lie on the margins of continents, and these are for the most part the areas in which large numbers of men have settled for longer time periods. It is estimated that over eighty percent of the population of eastern and southern Asia lives in such lowlands. The proportion of world population that lives in them is something on the order of seventy percent.

Moreover, from ancient times the seas have been the primary means for moving people and goods over long distances. Every continent is fringed by an array of port cities, large and small, which provide points of linkage with other areas—some of the cities along the same coasts, others across large intervening oceans and seas. In some parts of the world, as in Western Europe and much of Asia, certain of the port cities have grown to enormous size, and in numerous instances they are among the largest and most important metropoles in the countries to which they belong.

What is remarkable and unexpected, perhaps, is the apparently steady and seemingly inexorable drift of population toward the maritime frontiers, not only into the zone of two hundred miles distance from the sea or its lines of entry into the continents, but even within a few miles of the oceans—particularly along the fluctuating shores of the seas themselves.

The words "remarkable" and "unexpected" are used for several reasons. First of all, in Byron's time one would scarcely have made such a prediction. Men were preoccupied with the vast new territories which were just being opened to settlement, and the lure of the frontier was a powerful stimulus to men's imaginations. The Americas, particularly North America, had only begun to be settled by Europeans. Man's future in North America seemed to lie in the great interior of the continent although it was little known and understood as a geographical entity. South America and Africa seemed to offer equally great, though not immediately as attractive, opportunities. Moreover in Eurasia, the great continental interior—what Mackinder later termed the "Pivot Area" and the "Heartland"—also was being opened to Russian exploration and settlement. Somewhat earlier, Chinese power—under the Manchus—had directed its attention to the great interior, resulting in the creation of the "new province" of Hsin-chiang (Sinkiang) and in the confrontation of Chinese territorial and cultural imperialism with Russian expansionism.

Thus 150 years ago it would have been reasonable to assume that most of the world's expanding population would become concentrated in the interiors of the continents rather than along their margins, so that by this time in history the maritime provinces of the larger states at least would have become secondary to their continental cores. Of course knowledge concerning the geographical characterisitcs of the interiors was little understood. The western Middle West of the present United States was at once perceived as a Promised Land and the "Great American Desert." Moreover, the importance of the ports of entry into the interiors was clear enough, but theirs was seen as a linkage function rather than an otherwise creative one for large-scale occupancy and regional organization.

In addition it might have been reasonable to expect that with improvements in transportation technology, the importance of the seas as the main means for long-distance transport would have declined somewhat. The railroad in particular would seem to have marked the decline of other forms of transportation, at least in the great continental regions which seemed to possess the resource potentials for relative autarky. More recently, highway transport would seem to have moved in the same direction. Even fifty years ago it would have been reasonable to predict that only a few materials would have continued to move over long distances, fewer as time went on, and that, proportionally, maritime trade would have declined. The capstone in this evolution of new transportation relationships ought to have been the airplane because its advantages for high-speed movement have appeared superior to any alternative, and, in fact, it *has* come to carry an increasing share of people moving in the world—especially between continents where air transport now dominates.

Paradoxically, however, these postulates—though partially correct—have been disproven by the enormous vitality of maritime and other water transport on the world scene. Of course the characteristics of maritime trade have changed greatly in the past several decades, but the volume and the value of that trade has enormously increased and far more goods move between the continents than ever before. A world without shipping today is inconceivable. A country without access to the world ocean must be extraordinary in order to survive. Even in the United States, especially since World War II, it has been inland waterways linked with the coastal waterways system that has shown the highest rate of growth; and in Japan domestic cabotage has led the way to its economic "miracle."

These circumstances are the result of the increasing degree of specialization in production in the world. Indirectly, then, they are the result of an industrial revolution, which continues to transform the world pattern of economic relationships. Who in Byron's time could have predicted that Japan, a century and a half later, would have become the third most important industrial country in the world or, for that matter, even a few short decades ago? The increased dependence upon water transport especially for the carriage of bulky commodities and the economies of location associated with processing of these and even the further manufacturing of already processed raw materials which also move by sea have placed a high premium on tidewater locations. Industry means jobs and jobs mean people, multiplied of course by a constant relating to the power of such people to generate developments in the tertiary sector.

It is time however to back off from this argument and focus instead on the coastal zones of the world to see what areal similarities and differences there are among them, where the new types of ecological zonation have developed, and where problems are particularly acute.

In most of the world the interface between land and ocean is linear. It conforms for the most part with Byron's view. Man occupies the earth, for better or for worse, and his control and even his interest in the adjoining seas is small. This is true not only in areas of very sparse population, but also in areas of relatively dense populations. For example, along the Malabar Coast of peninsular India there is a comparatively narrow tidal zone of lagoons and offshore islands beyond which men seldom venture except for fishing from bamboo rafts, but even then they are seldom out of sight of land. Or again, along the east coast of Malaya where the northeast monsoon blows strongly on shore for four months of the year, man's "control" is negligible and his use of the territorial sea of his country is restricted to short forays—again for fishing offshore. Indeed, along most of the world's coastlines and certainly in the tropics, to the extent that man extends his interests and activities at all into the sea from land, it is primarily for fishing. In some instances, as on the coast of Chiang-su province in eastern China, near-shore fishing may be supplemented by partial reclamation for salt-pans; or as in the case of the northern Javanese coast, for fish ponds. On the whole, however, the interface is less a zone than a line; and the maritime component of it is particularly narrow, whatever national claims to territorial and continental shelf seas may be.

The opposite extreme from this condition is found in the industrialized countries, almost without exception. Consider the Eastern Seaboard of the United States, the site of the North American megalopolis. It is here that the interface becomes a zone of truly impressive dimensions, both on the land side and on the sea side; and the degree and intensity of interaction as occasioned by man, even apart from nature, is perhaps the highest in the world. The result is problems associated with the competition for scarce resources by a multiplicity of uses and with manifest disequilibria in what might be regarded as the "natural order" of relationships between land and sea.

Burton, Kates, and Snead [2] have divided the coastline from Virginia to Maine into four types of zones: the Village Shore, the Urban Shore, the Summer Shore, and the Empty Shore. The Village and Empty Shores involve most of the coast. They are sparsely settled and imbedded in their landscapes are the remains of what were once major fishing settlements, most of which have declined and left quasi-ghost towns in their wake. The maritime belt along these shores is narrow and much resembles that found in the nonindustrialized and urbanized regions of the world. The over-all zone, however, is relict in nature and is steadily declining in size as the other zones expand.

The Urban Shore consists of coasts devoted to urban types of occupance features. These areas are largely built up, and in them an industrial or industrially related type of land uses predominate. Obviously in this type of region are found the primary port facilities of the megalopolitan area, but almost all of these are associated with industrial establishments which find advantages to being on tidewater. More than half a century ago the primary locational consideration for these industrial establishments was proximity to cheap water transportation. Then, for a time, this variable became of lesser importance. In the last two decades it has again become of overwhelming importance, but there are other considerations. The ocean and its inlets provide an admirable source of water—for cooling in some cases, and for waste disposal in almost all. Thus, three locational considerations are fundamental to understanding these concentrations—water transport, waste disposal, and water as a coolant or even as a chemical in processing.

These considerations operate, however, within the context of the increasingly important locational pull of the market. That the area under discussion is part of the megalopolitan region means the existence of a market, but the concentration of industrial establishments also means the growth of that market—both as a matter of more or less "natural" forces and as the result of developmental policies both private and governmental. Here, then, is one of the major causes for the trend toward concentration of many people and kinds of activities in the coastal areas. However, it is not the only cause, and therefore it must be considered in relation to shifts in the raw-material sources for major industry in the United States toward overseas supplies of such materials as ores, petroleum, and even coal and salt, although these may also be carried in domestic cabotage.

The question of waste disposal does not relate simply to industries moreover, but obviously to the entire metropolitan complexes within which they are located. For cost reasons the dumping of solid waste takes place well within the shallower continental seas areas of the coast, and the effects this has on inshore fisheries and recreational uses of the entire coastal zone is the subject of much concern. The pollution of embayments and estuarial waters by this type of pollution is particularly acute; and it also has a major effect on coastal fisheries, especially those related to mariculture.

One of the principles for location of maricultural activities is that they be located near the markets for their products. This means the use of waters as near to the large urban centers as possible so as to reduce storage and transport costs, but it is precisely those areas that are most polluted. The impact of pollution on the fisheries of Chesapeake Bay and Long Island Sound, for example, needs no elaboration, and is apart from "over-fishing" as another element in the fisheries problem-complex.

The fourth type of zone, the Summer Shore, illustrates a major and compelling additional factor in the lure of tidewater and the creation of new eco-systemic relations in the coastal area. These are areas to which people move in large numbers during the summer months, chiefly from the nearer metropolitan areas, although not exclusively so. The recreational amenities of the coasts, not only in the northeast but also in most of the rest of the conterminous United States and Hawaii, provide a major attraction to an increasingly urbanized population. The Summer Zone overlaps with and is rapidly eating into the Village and Empty Shore zones, and it is probable that in most of the country the latter will disappear within the next few decades. Already an alarmingly high proportion of the coastlines of the country has been preempted in one way or another and barred to use by those who do not own land on the shore or beach, especially within easy driving distance of the larger cities.

In this situation there is basis for conflict both of perception and social purpose. The new compound coastal zone is characterized by competing types of occupancy, varieties of land uses, a heterogeneity of sea uses, all of them affecting the quality of the land-sea interfacial zone and often conflicting with what observers might regard as the "optimal use" of that zone.

In the United States the problems posed are by no means restricted to the Northeastern Seaboard, although the problems there appear to be particularly acute. West Coast states are similarly afflicted, but perhaps not to the same degree as yet. Moreover, similar kinds of problems characterize most of the Great Lakes. The issue of resolving contradictions in the development of coastal zones was never joined in the case of Lake Erie, the eutrophication of which is well advanced. It is now being joined with regard to the other lakes, especially Lake Michigan, but the elements of that issue have yet to be clearly identified. The problem of the management of the coastal zones of the Great Lakes is no less acute and as little touched upon as that of the management of the equivalent maritime coastal zones.

The United States is not, needless to say, the only country possessed of these problems. To a greater or lesser degree they afflict most of the industrialized states and will affect those now in process of industrializing. In Japan for example the rate of urbanization in recent years has been even greater than that in the United States, and now more than two-thirds of the Japanese live in urban places—mostly concentrated near tidewater. The Japanese have always regarded themselves as poorly endowed with natural resources, including land in the broadest sense. Their perception of how land should be used (for various reasons too complex to go into here) has militated against their converting any more paddy land than absolutely necessary into urban land uses; and they have been willing, at least thus far, to pay some of the costs for such a policy.

The alternative has been to reclaim land from the sea, especially in shallower waters of the alluviating bays that mark the southern and urbanized coastal zone of Japan. Thus, Tokyo Bay is now almost entirely ringed with industrial land uses on reclaimed land, and the same is true of the shores of Osaka and Ise Bays. Virtually no land has been retained for recreational purposes, to the disadvantage of the burgeoning populations of Japan's megalopolis, from Tokyo to Okayama on the Inland Sea. Of course, land reclamation has a long history in Japan. Much of the best paddy land in Japan was reclaimed in pre-modern Tokugawa times. The solution to the land problem as seen by the Japanese has been in keeping with Japanese traditional culture, although the new wave of massive reclamation has been caused by nontraditional stimuli.

Not only has the shoreline of much of southern Japan been modified by these developments but new problems have arisen, one of the more important of which is that of land subsidence. Whereas land could be reclaimed for industrial purposes, it needs a long period to settle. Economic motivations have prevented taking the necessary time for the land to settle before construction started. Moreover the low-lying reclaimed areas are inevitably vulnerable to high tides, tidal waves, *tsunami*, and other forms of oceanic disturbance. Thus, tidal walls and gates have been integral parts of the reclamation programs. Land has subsided, however, at such a rate that these have been damaged and have to be rebuilt or replaced. The prime reasons are haste in building (as noted) and the use of sub-surface water for industrial purposes through wells sunk on the properties concerned, a factor related to the acute shortage of water supplies for the larger Japanese cities. The result has been subsidence significant enough to threaten large areas of industrial development in the tidewater portions of Japan's industrial belt. As a result government and industry are confronted with the question of who pays for damages and for the construction of new flood protection works. The issue is one of the more important in Japan.

At the same time inshore waters have been used for sewage disposal. The waters of Tokyo Bay have become so polluted that dumping of urban waste in the bay has been forbidden. Nonetheless, the shoreline has been preempted to such a degree that fishermen could not operate in the bay even if its waters were less polluted. The most affected are shellfish and seaweed farms, for which location near the market areas would be vastly advantageous. The governmental response to this problem has been to buy off the fishermen, most of whom then move to the city proper or its suburbs and change their occupations.

The Japanese case may be exacerbated by the constraints placed upon it by natural conditions, but it contains points in common with most such instances of the creation of a new coastal environmental zone composed of land and sea elements as employed and modified by men. Academia appears to be inundated by a plethora of "research frontiers" along which major research efforts should be directed. It is of note that the coastal zones in transformation are not only such a frontier, but also one of markedly high priority in the scientific scheme of things. The problem is man's use not only of the earth but also of the adjoining sea, and the ways in which he has integrated both into new subeco-systems—the nature of which are as yet poorly understood.

WENDELL A. MORDY and BROOKE D. MORDY

4 Atmospheric Control and the Ocean Regime

Large-scale weather and climate control has been set as a long-range goal of scientific research by the meteorological profession, nationally and internationally, by the United States and other governments, and by the United Nations General Assembly. Meanwhile, it is recognized that inadvertent modification is occurring due to urban and other construction, transport, irrigation, deforestation and agriculture, but the significance and extent of this has not as yet been determined.

The growing scale of unintentional modification of climate, much of it possibly detrimental to life, already makes new institutions imperative—institutions designed to control the anticipated problems in a global context. Furthermore, some types of intentional large-scale weather manipulation would be technologically possible today; but to even consider, much less attempt, such undertakings, new institutions are first needed. The proposed Ocean Regime, conceived as governing the use of ocean space—which of course includes the atmosphere, may at least in part offer an answer to these needs.

Some of the proposals for changing the climate on the largest scale involve altering certain characteristics of the oceans. Conversely, super-scale hydroelectric and irrigation projects are being

WENDELL A. MORDY is the Director of the Sea Grant Program, School of Maritime and Atmospheric Sciences, University of Miami.

BROOKE D. MORDY was educated as an anthropologist and sociologist. She has taught and done research in social anthropology.

discussed which quite possibly might change the circulation of the atmosphere in such a way that the oceans would be affected. Also, new techniques which will be developed for exploiting ocean resources may be expected to have significant influences on the atmosphere.

Although the interrelationships in the oceans and atmosphere are extremely complex and incompletely understood, the above statements serve to underscore the importance of regarding them as a single system, a point of view fundamental to contemplating controlled development in either medium. A recent sign of the scientific recognition of this is that a new journal, the *Journal of Physical Oceanography*, is to be published by the American Meteorological Society beginning this year. A quarterly, it will be devoted to "the communication of knowledge concerning the physics and chemistry of the oceans and of the processes coupling the sea to the atmosphere."

Focusing on the atmosphere, two other points are basic to a realistic outlook as man continues to modify the atmosphere through various activities and to move toward larger-scale deliberate control. First, if man chooses to change climates or even to control storms, he will be controlling many orders of magnitude more energy than ever before in history. Second, because all life is weather related and climate dependent, man doubtless will be profoundly influencing life patterns on a global scale as he induces climatic changes.

The Weather Working Group of the Ecological Society of America said in a report to the NSF Special Commission on Weather Modification:

> Present concern about weather modifications stems from more or less imminent operations on systems of intermediate scale . . . (and) it is likely that the biological consequences of intermediate-scale operations will be very serious. . . .
>
> Existing theory relating organisms and their climatic environment is far from being adequate to meet the demands that these operations are likely to place upon it.

As for large-scale modification, the report stresses the potential dangers to the eco-system and advocates extreme caution.

> Widespread extinction, complete community reorganization, disruption of ecological systems, and serious outbreaks of pests, weeds, and disease organisms would almost certainly follow any sudden and severe modification of the global climate. Under present circumstances, we can only regard the application of global-scale weather modification as an unmitigated biological disaster and hope that we may attain some better understanding of how to handle such a genie before geophysicists are able to uncork the bottle that holds it.[1]

Fortunately, weather control is the first case in which the development of a major potential technology is being accompanied by attention to the multiple implications of its application. Thought is already being given to the probable ecological, social, economic, political and legal ramifications of control, national and international, and caution is generally advocated.

Gilbert White said in 1965 that "the reasonable fear that a modification in a small area might have profound impacts far away" acts as an "incentive to investigation of the human dimensions of weather modification and makes essential a broad front of study sensitive to local, national, and international implications of whatever chain of events may be generated in the interlocking systems of air, water, plants, animals, and man." [2]

INTERNATIONAL ATMOSPHERIC MODIFICATION AND THE OCEANS

As already noted, man is modifying the weather—unintentionally. Also, means to influence the weather exist now, and further means are certain to be developed. The ultimate limitations on the application of weather control techniques will not be technological, but rather political.

Some confusion seems to persist on the question of whether or not the weather can be intentionally changed. This is probably caused by the controversy surrounding the evaluation of field experiments testing the effectiveness of cloud seeding in producing rain or snow—only one of several ways in which men are now influencing or propose to influence the weather and climate but the way which has received the most public attention. Techniques exist and sufficient energy could be commanded to change weather in several ways, both on a small scale and on a large scale.

Present capability for manipulating the weather can be categorized into modifications which can be effected within the atmosphere, and changes which could be brought about by manipulation of the ocean or of land masses. We will return later to the first and more familiar type.

The latter type of modification is not taken as seriously, because while it is possible in an engineering sense, it is, so far, not contemplated as prudent. The technical feasibility of this type, however, can be easily demonstrated. It would be technically possible to begin a project tomorrow on the island of Oahu which would with virtual certainty result in reducing the rainfall in Manoa Valley by at least one-third. This example raises the questions and problems which ultimately must be faced in the practice of all weather modification.

In addition to suggestions for changing mountain contours, drying existing inland seas or creating new ones, and diverting the Kuroshio, Gulf Stream, and similar climate-dominating currents, there are proposals to influence the radiative balance of the earth by altering the extent of ice in the Arctic Ocean, or by placing sunlight-diverting obstacles in space.

Variations in the amount of cloud, or in the extent of snow and ice, vary the amount of heat available to drive the atmosphere and oceans because these surfaces reflect a large percentage of incoming sunlight back to space, and reflected radiation does not appreciably warm the atmosphere. Reduction in average cloud cover results, among other things, (ignoring, for present purposes, all the complexities) in more evaporation and thus subsequently in the formation of more sun-reflecting cloud, in a rather promptly self-regulating system.

Ice extent, however, probably does not have the rapidly self-regulating stability of cloud cover. It is known to have varied greatly in the history of the earth, and it is likely that different pack conditions were accompanied by climatic variations. Therefore, an often suggested means for altering the radiative balance of the earth is to change the extent of Arctic Ice. A number of ways to do this have been suggested.

One scientist has suggested damming the Bering Strait, then pumping Arctic Ocean water into the Pacific which would cause warm Atlantic water to flow in and melt the ice. Another has proposed that atomic-bomb explosions could produce persistent ice fogs which would allow the ice beneath them to melt by slowing down the escape of heat into space. Others have suggested that the energy to melt the ice could be made available by blackening the surfaces.

Some scientific opinion holds that if the Arctic Ocean were once cleared of ice, it would remain free of it. Others suggest that additional heat would be required to keep the ocean open, heat which could be supplied by providing for exchange of water with the Atlantic or Pacific.

No one can say with assurance what the result of such undertakings might be in regard to the general circulation of the atmosphere. In fact, it hardly need be pointed out that the atmosphere is not well enough understood as it is, for forecasts to be made with certainty. On a large scale prediction must precede even the contemplation of control.

However, a less positive view of the potential than the one expressed above was elicited by a team of American meteorologists visiting the Soviet Arctic and Antarctic Institute recently. (Reported by L. J. Battan in the December, 1969, Bulletin of the American Meteorological Society.)[3] The Russian scientists felt that much more must be learned about the Arctic before claims about climatic change can be made, and that ideas such as the Bering Strait scheme have no merit.

Other suggestions, not safely to be dismissed as sheer fantasy, include the creation of a Saturn-like ring of potassium dust around the planet, which according to calculations should produce a twelve percent increase in solar radiation incident on the earth. The purpose of such projects again would be to reduce or eliminate the ice caps and ameliorate arctic and sub-arctic climates. Another means suggested is the creation of "artificial suns"—reflectors in space. Both ideas would effectively turn night into day.

While these projects seem unthinkable, it may be recalled that the United States has already exploded an atom bomb in space and dispersed an electromagnetic wave reflecting dipole antenna wires in orbit around the earth, unilaterally and against the cautionary advice of many scientists.

Let us return to weather modification as it is commonly referred to—that is, triggering of changes by manipulations within the atmosphere, principally by cloud seeding. A major part of this research has been devoted to efforts to produce snow or rain, dissipate fog, or steer hurricanes by directly intervening in the cloud and fog forming process. Some practical as well as scientific progress has been achieved in such work, but the generally accepted or uncontested results involve clouds and fogs on a scale of only a few square miles in suitable meteorological conditions.

While effects have been observed, cloud-seeding procedures affect only one of many parameters of the cloud-forming, thermally driven, atmospheric machine—the whole of which is insufficiently understood to allow easy, or in most cases, reliable evaluation of the results of interference. Of comparable importance to the man-induced cloud seeding effects are the naturally occurring influences on cloud and raindrop or snowflake formation, such as the concentrations and sizes of nuclei on which the cloud droplets condense, the rates of cooling and condensation, and the naturally occurring clay fragments and other particles found in the atmosphere which can also produce incipient ice crystals in clouds. The motions of the cloud are governed not only by the release of energy in the condensation and freezing of cloud droplets but also by the stratification and wind structure of the surrounding air and by the topography below.

Whatever the method contemplated for large-scale weather modification, it should not be undertaken without scientific testing and accurate advance knowledge of the results. This requires sufficiently precise knowledge of the normal behavior of the atmosphere by theoretical meteorology that the intended result of a proposed modification can be predicted with confidence. Because of anticipated widespread consequences of changing large-scale weather phenomena, it would seem that certainty beyond that needed for weather prediction service would be required.

Thus, the research needed both for weather prediction and for control converge in the objective of developing precise theoretical models of the atmosphere, which can be tested by computing atmospheric changes in advance, and then comparing computed behavior with subsequent actual atmospheric behavior.

Hypothetical experiments in weather modification are also contemplated to explore, with the use of large computers, the effects which might result if experiments were in fact undertaken—the only safe way to proceed.

While the tempo of theoretical work and data gathering is increasing with the programs of the Global Atmospheric Research Program and the World Weather Watch, the science of meteorology is still far from its objectives of not only observing but also understanding the atmosphere in sufficient detail. Rapid progress is occurring in observation through international cooperation in the employment of weather satellites and other new technologies.

The more knowledge and understanding of weather processes increase, the more advisable it would seem that the proposed Ocean Regime have the capability of monitoring the weather in ocean space, so as to be in a position to detect undesirable modifications with an impact on the oceans.

UNINTENTIONAL ATMOSPHERIC MODIFICATION AND THE OCEANS

In the realm of inadvertent weather and climate modification, man has been active since he first wielded an axe or put a plow into the soil. Life forms depend on climate, but also affect it. Anyone who has run barefoot knows the differences in temperature experienced as one walks over packed soil, grassy fields, plowed ground, or paved surfaces. A significant amount of the earth's surface has been altered by timbering, man-made forest fires, agriculture and irrigation, urban-industrial development, and road construction. These activities also involve waste disposal to the atmosphere.

Mostly, the effects of man's activities on the world's weather are unmeasured, but some are clear. Clear also is the fact that they are increasing—many believe to an alarming extent.

For example, city climates are up to five degrees centigrade warmer than the surrounding countryside. Cities serve as heat sources which increase convection and cloud formation and also form barriers to normal wind flow. They are a source of large quantities of particulates and pollutant materials in the atmosphere. Some of these serve as nuclei upon which droplets and ice crystals form in clouds. Rainfall amounts downwind of cities increase as the cities grow; up to seven percent increases in rainfall have been determined in studies. Spots in the sea, warmed or contaminated by activities on the sea floor below, in future, may prove to produce similar meteorological influences.

Very recently satellite photographs have revealed urban center influences on marine cloud patterns leeward from the east coasts of America and Asia, particularly New York, Boston, Philadelphia, and Tokyo. In outbreaks of cold air from the land to the sea, fingers of cloud reach out from these cities into the cloud banks lying thirty to fifty miles at sea, influences which are recognizable for five hundred miles or more within the cloud bank. The relative importance of factors such as pollutants, heat, and water vapor contributing to these clouds is unknown, as is the geographic extent of their current influence on weather. It may be presumed, however, that as world industrial development increases such influences will also increase.

In the central subtropical Pacific, far from continental sources, particulate matter in the atmosphere as measured from nearly fourteen thousand feet on the top of Mauna Loa, has more than doubled in fifteen years.[4] Its specific source is unknown, but there is little doubt that it is related to man's activities. Such small particles have a residence time in the atmosphere sufficient for them to circle the globe more than once. Thus in their role as cloud droplet nuclei, they have a potential for global influence.

Noticed and discussed with increasing interest are clouds and fogs caused by human activity. As yet their meteorological significance on a world scale may be minimal but they are noticeable in photographs which encompass the entire sunlit hemisphere, taken by the weather satellites located eighteen thousand miles out in space. It has been observed on occasions that a single ship will produce in its wake a fog or a low-stratus cloud deck covering an area of six thousand square miles, formed in the smoke and water vapor the ship adds to an already humid but very clean environment. In such meteorological conditions, as many as eight ships may be observed leaving fog or cloud wakes, increasing by many times the area so covered. Until lapse-time satellite cinephotographs revealed this ship-produced phenomenon, meteorologists were unaware of the man-induced origin of such clouds, for they form well behind the ships and are indistinguishable from other fog or clouds at close range.

Man-made clouds produced in the upper atmosphere are being discussed with increasing frequency as a potential for *large-scale*, inadvertent weather modification. The satellite photographs mentioned above reveal natural phenomena which suggest this possibility. For example, on occasions when thunderstorms are prevalent over Florida, moisture is carried sufficiently high in the atmosphere to freeze easily into ice crystals. Cirrus, or ice clouds, formed in the anvil-shaped tops of many cumulo-nimbus clouds, mingle, and show in the photographs as huge horizontal shields of ice particles. These ice particle clouds drift out over the Atlantic Ocean where they appear to "seed," with their naturally produced ice crystals, marine cumulus clouds which penetrate them from below, thus increasing the potential for rain in those areas affected by the cirrus. Similar striking examples occur near the Yucatan Peninsula where winds carry cloud-seeding cirrus into land-produced cumulus clouds in Mexico to the south.

The relevance of these observations for human influence is that jet aircraft are also known to be frequent producers of wide areas of cirrus clouds. Depending on the persistence and trajectories of these clouds it is not only conceivable but probable that on occasion they play a role in precipitation formation over broad areas. Jet aircraft also add radiation *absorbing* water vapor (selectively) to the stratosphere, and the radiation *reflecting* properties of the clouds they produce is being studied for the effects such sunlight-shading clouds may have on the weather. The heavily traveled New York to Chicago route is known to be cloudier today than before the advent of jet travel.

Hurricanes and typhoons create areas of cirrus clouds hundreds of thousands of square miles in area. Seeding hurricanes would among other influences have the effect of producing more cirrus clouds, and at lower levels—increasing the likelihood of subsequent meteorological influences some distance away. Hurricane "Candy" recently produced a cirrus cloud shield two hundred miles wide and six hundred miles long.

Other possible weather-influencing, man-produced effects, visible eighteen thousand miles out in space, include forest fires and burning oil wells. An oil well burning near Los Angeles last year produced a white plume of either cloud or smoke two hundred miles long.

Another possible inadvertent influence on the weather has come to light recently. Lead iodide, like silver iodide mentioned earlier in connection with cloud seeding, is a very good cloud-seeding substance. Recent discoveries indicate that substantial amounts of lead iodide are formed in the atmosphere as a result of the combination of lead, released in burning automobile fuel, with iodine from natural and industrial sources. The importance or extent of this influence has not yet been determined. However, because the particles are extremely small and suspended for long duration in the atmosphere, such pollutants could easily be globally pervasive.

Now that men are moving into the sea with similar intent to that with which they once moved into the forests or settled the land, some questions relevant to atmospheric protection and development will continually arise. What kind of projected activities in the sea should be monitored or regulated because of potential influences on the atmosphere? What kind of technical data should be sought to prevent unanticipated harmful consequences? What interactions between sea and atmosphere may be affected by human activities?

INTERACTIONS BETWEEN OCEAN AND ATMOSPHERE

Major interactions between the ocean and atmosphere occur at their interface, where energy is absorbed and exchanged. Special attention should be paid to activities in the ocean which may alter the temperature or the chemistry of the sea surface.

Breaking waves and bursting bubbles cast sea water and fragments of surface film into the atmosphere. Droplets, evaporated down to their solid nuclei of salt and organic substances, together with film fragments, are carried aloft by winds. Conversely, solid material within the atmosphere is returned to the sea surface by settling or is coagulated by cloud processes and washed out by precipitation. Sea-originating particles play an important role in the formation of cloud and precipitation and in the nutrient cycle for plant life on land. Here at the sea surface gases are also exchanged between the air and water which are essential to life.

The subtleties of the thermal exchanges between ocean and atmosphere are recognized but not well understood. Those who have used hot water bottles are aware of the heat storage capacity of water. Thus heat sufficient to increase the temperature of the entire atmosphere by six degrees centigrade would produce only 0.015° C. temperature change in sea water to a depth of one thousand meters. Currents running deep in the ocean and insulated and isolated from surface weather for hundreds or possibly thousands of years are believed by some to contain temperature histories of the past. Upon ultimate surfacing, the currents are believed to produce secular changes in weather and climate. There is at present no adequate way to test such an hypothesis. Thus if heat-dispersing atomic power plants are planned for construction on the sea bed or sea surface in numbers or concentration sufficient to alter ocean circulation, it is not inconceivable that weather in their vicinity or in a wider area might be influenced.

The most important influence of expanded sea bed industrialization, from a meteorological viewpoint, would probably be the effect on the surface layers of waste disposal and sea floor agitation or of upsets in the biotic balances as a result of such activities. Thin films of biologically produced material only a molecule or two thick can inhibit evaporation, increase the temperature of surface waters, or both. Surface films could not only influence the sizes of sea-produced particulate matter in the atmosphere, but conceivably they could also become a new or increasing source of air pollution.

It is not improbable that in the decades to come, human activity under or on the sea will produce surface-active layers over extended areas sufficient to affect cloud nuclei production, alter evaporation, or displace sea life. The entire sea could be covered by smaller amounts of such materials than the amounts of other air-borne particulate matter now circling the earth produced by land-based human activities.

Much more has been written about schemes to alter weather, and about man's unwitting alteration of it. The foregoing discussion is intended to emphasize that modification has occurred unknowingly, that it can be projected on a larger scale, and that such modification involves both the ocean and the atmosphere.

THE CONTROL OF ATMOSPHERIC CONTROL

The process of defining and classifying the atmosphere from the points of view of law and economics has begun, and will continue to accompany changes in atmospheric uses. Apparently no adequate body of law exists, even domestically, to deal with problems which arise as we begin to use the atmosphere as a manageable resource. Howard Taubenfeld, in a brief paper pointing to the virgin field of weather modification for the international lawyer, says in regard to weather control: "If we look only at potential international problems, our machinery for decisionmaking ranges from primitive to nonexistent." [5]

Pointing out that "the doctrine of *res nullius* is a doctrine applicable to the free development of a resource as essentially a free good," Vincent Ostrom, writing on the need for research on the political aspects of the human use of the atmosphere, says:

> Much of the contemporary law regarding the use and development of the atmosphere has not advanced beyond the conceptions inherent in the doctrine of *res nullius*.[6]

The new ideas about social property being discussed as appropriate for exploration and development of the high seas certainly seem to be more adequate than earlier concepts to the needs for the ocean, and they appear to be equally applicable to the atmosphere above the seas. In fact, considering the constant unbounded motion of the global atmosphere, a single administration in charge of its exploitation for the good of all mankind is certainly a desirable goal in the long run. But of course, as Rita and Howard Taubenfeld say in a recent article considering some of the international implications of weather modification: "States . . . claim the airspace above the national territory and waters and seem certain to assert rights of 'ownership' or control in the clouds and other weather phenomena in national airspace."[7]

But the airspace over the high seas is not subject to national claims, and regulation of it by a responsible international body would certainly seem to be a long step in the right direction. And it would be appropriate that such a body also be concerned with the oceans, for the oceans would be crucially involved in any future plans for large-scale weather control and protection of the atmosphere.

While ocean space by definition in the draft statute includes the atmosphere above the high seas, territorial waters, and contiguous zones, the statute is primarily aimed at uses of the high seas and sea bed, and it is not always clear to what extent the atmosphere was intended, or could be interpreted, to be included in its provisions. (See Appendix p. 331.) The objectives of the Regime for development, conservation, and security of the high seas and the sea bed can only be met if the interdependency of ocean and atmosphere is fully taken into account. Perhaps more explicit provisions should be included regarding the atmosphere than now appear in the draft statute.

The long history of international cooperation in meteorological research, and the official and professional organizations which exist to facilitate it, could be assets to the proposed new Regime. And of course meteorology and oceanography, closely allied disciplines, are already working together on international programs of observation and research.

Thus meteorology has something to offer the Ocean Regime, and conversely, the Ocean Regime may be a first answer to the widely felt need in the field of meteorology and among interested statesmen and lawyers for new institutions to deal with emerging problems of atmospheric management. For the potential of large-scale weather control and the expectation of increasing modification as a side effect of other activities together point to an eventual if not an immediate need for worldwide atmospheric resource management.

Of course, as is true for all powerful technologies, applications can be constructive or destructive, and for years many have warned of the potential strategic capability of weather modification. A number of responsible scientists, including John Von Neumann, Irving Langmuir, and Thomas F. Malone, have considered it an ultimate threat as a weapon. Rita and Howard Taubenfeld, in a recent article considering some of the international implications of weather modification, point out that its potential strategic capability is an inescapable factor underlying all considerations of achieving international accord on its use.

They observe that, as long as modification is on a small scale and potential harm is minor, the existence of formally organized cooperation can facilitate peaceful use. Voluntary mutual accommodation is more likely where there are already patterns of international cooperation.

The Taubenfelds believe that: "The development of organized international cooperation concerned with operational weather modification activities of a . . . type (such as) fog dissipation, rain augmentation, hail and lightning suppression, or even hurricane steering . . . may perhaps be expected." But this "would leave the states free to pursue their own competitive research programs including those relating to major weather modification," or in other words research activities with strategic applications. The Taubenfelds point out that: "Major climate changes might well affect the eco-system and the health, mores, food, shelter, livelihood, and way of life of vast areas of the world. It could precipitate either weather retaliation or major population migration." This would inevitably lead to international conflicts, and "these are the types of fundamental major political conflicts that cannot normally be successfully subjected to legal conflict resolution or to organized technical cooperation and that have not historically been settled peacefully in the present international system."

In further discussion, they stated that:

> . . . the traumatic potentialities of major weather switching suggest two interrelated policies for states. First of all, . . . all major powers have strong incentives to compete in the development of weather modification technology. An arms race psychology to technological developments in this field alone appears safe and "rational." Second, since major institutional changes in the present international system

> cannot be expected to be readily achieved, at least those nations with relatively good weather also have incentives to prevent or delay the effective development of weather modification capacity of the major weather-switching variety so long as they can do so safely—i.e., without running the risk of being bypassed in the weather modification knowledge race.[8]

They conclude that in the long run, control of large-scale weather modification technology probably implies "major institutional modification of the international system."

The proposed Ocean Regime would provide a new kind of international institution more adequate than those now in existence for achieving progress toward the goal of peaceful and rational use of the atmosphere. The Regime could serve to further institutionalize the cooperative international research efforts already underway, and it would provide a deliberative body and a mechanism for planning for atmospheric conservation and weather and climate control over a large part of the globe. The World Meteorological Organization serves well, but has been limited to encouraging research and the exchange of information and providing a worldwide observational network.

An international body governing more than seventy percent of the atmosphere, dedicated to peaceful uses of that space, would be an influence for the development of constructive rather than destructive applications of emerging weather control technology.

Political feasibility aside, it might even be suggested that, considering the unbounded nature of the weather and the unity of the physical and ecological system of the ocean and all the atmosphere, the Ocean Regime should have explicit responsibility for rendering "recommendations and opinions" affecting the total global atmosphere. This might be by extension of the provision that the Regime "render recommendations and opinions affecting areas under the jurisdiction of coastal States or island States." Whether such a provision would be politically advisable in regard to initial acceptance of the Ocean Regime is no doubt a question for others to consider.

At any given time there is a limit to the tools we possess; there is a limit to the scientific understanding of a given phenomenon; there is a limit to our capacity to comprehend the complexity of the system or systems in which the phenomenon is imbedded, and therefore a limit to our ability to predict the implications for the system of a change in one part of it. In other words, we may have the technical ability to effect changes, but "control" will in no sense be absolute—we will not be able to anticipate all consequences, and we may not be able to reverse processes once set in motion if we decide that they are undesirable.

In the case of weather and climate, involving larger forces than man has yet manipulated, there seems to be consensus among those who have considered these problems that nothing irreversible should be attempted. Uncertainty characterizes our understanding. Weather modification involves alterations not only in the atmosphere (climate is by nature inconstant) but also in the systems that interlock with it—hydrological, biological and cultural. No one of them is fully known, so the results of modification cannot all be identified, much less predicted.

Ethical considerations are profoundly involved in weather modification. The goal of climate control is generally couched in terms of the benefit of all mankind. It is not surprising that men aspire to control the threats to their lives from hurricanes, blizzards, typhoons, droughts and floods. Nor is it surprising that they would aspire to make the desert or the frozen tundra bloom. But large-scale control will inevitably involve large numbers of people—people no doubt representing a diversity of values and interests. Where there are differences, questions of human rights must be resolved.

Ultimately the world needs a responsible body acting for the welfare of mankind as a whole which has legal and political authority for control of weather control, and of all atmospheric modification. Meanwhile, the Ocean Regime may offer hope of partially filling this need.

SUMMARY

The implications of weather modification and control for an Ocean Regime were considered. The importance of regarding the ocean and the atmosphere as a single system, global in scope, was emphasized. Some evidence of increasing inadvertent modification of the atmosphere through activities on land, and in the sea and the air, was discussed, and the prospect of scientific capability for atmospheric control was considered.

The recognition that these developments and eventualities have serious ecological, social, economic, legal and political implications was pointed out. Because climate control would involve manipulating many orders of magnitude more energy than ever before, which undoubtedly would influence life patterns on a global scale, caution is urged.

Acknowledgment was made of the fact that international cooperation in meteorological research and service, both official and professional, is outstanding and characterized by foresight. Nevertheless, it is suggested that new international institutions may be needed to supplement existing arrangements and to control the anticipated problems and developments in a global context.

WARREN S. WOOSTER

5 Oceanography and International Ocean Affairs

International ocean affairs may be defined as the complex of interactions among nations and their policies with respect to the use of the ocean and its resources and to the governance of that use. National ocean policies may on occasion have been deliberately and logically developed, but more commonly they have arisen as conglomerates of assorted positions and practices. Ocean policy reflects national political interest which is based in turn on economic security, social and philosophical considerations. Internationally, such policies properly go beyond immediate national political interests in their concern with the reduction of political conflict between nations and with the conservation and rational use of marine resources as a common heritage of man.

Ocean policy has not been derived from ocean science, nor has ocean use depended in a direct and major way on scientific knowledge. The massive shipping industry has developed over centuries with scant assistance from the oceanographer. Historically, the major fisheries have arisen without much, if any, of a scientific base. Only when submarine and antisubmarine warfare became important did navies first become strongly interested in oceanography.

The ocean user normally has no more than the layman's curiosity about how the ocean functions. He is concerned with using the ocean and its resources, either for profit or for a number of

WARREN S. WOOSTER is a professor at Scripps Institution of Oceanography, University of California, San Diego.

other socio-economic or political reasons. Ocean science is of interest if it improves his ability to conduct his affairs at sea.

But the effective use of the ocean and its resources is rapidly becoming more dependent on ocean science. The petroleum industry has required geological and geophysical studies for the development of resources on the continental shelf. Exploitation of deep-sea minerals will only be possible at an economic level if scientific studies of the potential resources are carried out. The fishing industry requires scientific knowledge if the living resources are to be maintained and harvested at optimum levels. The use of the ocean as a receptacle for wastes must be based on knowledge of its capacity for such materials. And weather forecasts of long duration will depend on better understanding of the upper layers of the ocean.

Thus ocean policy will to an increasing extent require a scientific base, particularly in the selection of alternative solutions for resource and security problems. In view of this growing dependence, the way in which the necessary scientific information is acquired is relevant to the consideration of international ocean affairs.

Ocean research is conducted in many countries by individual scientists concerned with describing and understanding ocean processes and phenomena. What is the professional interest of these oceanographers in the development of international ocean affairs? Their primary interest is in those aspects that serve to promote scientific investigation of the ocean. Because of their specialized knowledge, they are also interested in actions that tend to increase the benefits to mankind from the ocean. Finally, they are interested in influencing ocean policy as it affects either the scientific investigation or the effective use of the ocean. The special interest and competence of oceanographers can be assumed to decrease in the above order as the scientific or technical component of the problem decreases.

This paper is concerned with the following aspects of the relation between oceanographers and international ocean affairs:

1. In what ways can international action further the scientific investigation of the ocean and its resources?

2. What mechanisms exist whereby such international action can be carried out?
3. In what ways should intergovernmental policy be shaped to further the scientific investigation and effective use of the ocean?

INTERNATIONAL COOPERATION IN OCEANOGRAPHY

In two general areas of ocean research, positive and active international cooperation is required for significant gains in understanding to be achieved. These are the studies of processes of ocean dimensions and those of widespread ocean resources. In the first case, such studies often require international cooperative investigation because of the vast scale involved and the relative slowness with which observations can be made. These processes operate across national boundaries, and their understanding is important to people in many countries. In the second case, international cooperation is required to establish an agreed basis for rational international use of the resources.

In all fields of oceanography, processes have been observed that operate over vast areas of the ocean without regard to the arbitrary boundaries drawn by man. For example, much of the planet's weather is determined by the exchange of heat between the lower atmosphere and the upper layers of the ocean extending across five-sevenths of the earth's surface. The waters of the ocean move endlessly and are interconnected so that an observed volume of water may at some previous time have been almost anywhere else in the ocean. The processes that determine the dissolved substances in sea water are seldom restricted to a local area. The structure of the sea floor reflects forces that are reshaping the entire planet. Marine organisms are so interlinked that an element absorbed in the food web at one location may reappear months later and thousands of miles away.

To some extent, studies of widespread ocean resources incorporate all of ocean research. The spectrum of possibilities extends from the most fundamental studies of the structure of water to the most applied development of fishing techniques. But some investigations are directly relevant to resource utilization and depend on international action for their success. These include: (1) exploration and assessment of resources, (2) the study of processes affecting the abundance, distribution, and availability of resources, (3) studies related to resource extraction, and (4) studies required for management and conservation of resources.

For example, potential nonliving resources of the continental slopes and deep ocean floor cannot be evaluated within a reasonable period of time unless the countries concerned pool their exploration efforts and information. Studies of processes affecting the abundance, distribution, and availability of these resources are required as a guide to prospecting and detailed reconnaissance. Intelligent decisions on the rules, principles and procedures for the rational use of these potential resources cannot be made by the world community until the essential information is available.

Proposals for international cooperative scientific investigations are often initiated by individual scientists or groups of scientists, whereas proposals in applied science usually arise directly from governments, often in response to social and economic needs. Programs of both types are carried out by various combinations of laboratories and institutions. Experience has shown that successful cooperative scientific programs depend on the interest and involvement of scientists from the earliest stages of planning. It also seems to be true that the success of such programs depends on coordination being kept to the least formal level compatible with the complexity of the program in question. Thus, where possible, informal agreements or those at a regional level are preferable, whereas formal intergovernmental planning at the global level introduces complexities that are only justified when the investigation is one of truly global character.

MECHANISMS FOR INTERNATIONAL ACTION

In considering mechanisms for promoting scientific investigation of the ocean and its resources, it should be noted that such investigations are composed of the efforts of individual scientists. The needs of these scientists at the national level include financial support, basic facilities (library, laboratories, equipment, ships), and scientific assistance and services. When participating in cooperative investigations, further needs are ready communication with other scientists, and the services of a planning and coordination mechanism.

Based on these requirements, the following kinds of mechanisms are considered:

1. Funding mechanisms
2. Planning and coordination mechanisms
3. Communication mechanisms

Funding mechanisms

In many countries, oceanographic research is carried out in both university and governmental laboratories. Because of the expense of such research, funds are generally obtained from public sources. To a small extent, academic research may be privately supported, but from the international point of view private funding can largely be ignored. Private ocean research is also conducted by industry, most notably by the petroleum industry, but for the most part the results of such research are proprietary and unavailable to the international community.

Within the government, ocean research programs may be found in a variety of operating agencies such as those concerned with fisheries, with national security, with charting and other maritime services, with weather monitoring and forecasting, and with the development of mineral resources. Research programs of such agencies are usually conducted in their own laboratories, although in some cases the work may be contracted out to academic laboratories or to industry. In addition, mechanisms often exist for the specific purpose of channeling governmental funds into academic laboratories (such as the National Science Foundation in the United States or the University Grants Commission in the United Kingdom).

On the international level, funding mechanisms serve a different purpose, since the principal expenses are met nationally. International public organizations are more concerned with providing technical assistance to developing countries, largely in connection with the development of marine resources.

Some general comments on the relative roles of developed and developing countries in international cooperative investigations should be made here. Only a few industrialized countries have a major capability in large-scale modern oceanographic research. Only these countries have a sufficient number of trained marine scientists and ample funds and facilities for carrying out such research. And of these countries an even smaller number has a reserve of oceanographic effort that can be applied to research in distant waters. The smaller advanced countries devote their efforts to work in the vicinity of their coasts or in nearby seas.

It is important for developing coastal countries to achieve their own competence in ocean research and to participate actively in investigations off their coasts.

Largely in response to the needs of developing countries, international operating agencies such as the United Nations Educational, Scientific, and Cultural Organization, the Food and Agriculture Organization and the World Meteorological Organization have relatively large secretariats and budgets and programs of their own. However, only a small fraction of the efforts of such agencies is devoted to activities contributing directly to ocean science. Larger amounts are available for resource development through the United Nations Development Fund, under which the UN organizations serve as administering agencies. But only a small part of the Special Fund programs is allocated to scientific investigation of the ocean, as contrasted with the much larger component of technological, economic, and training aspects.

An important contribution of UNESCO to science in general as well as to marine science should be recognized here. For some years UNESCO has been a major source of income to the International Council of Scientific Unions and its constituent bodies including the Scientific Committee on Oceanic Research. Although the sums are not large relative to the cost of doing research, they have been an invaluable help in the functions of scientific communication and to some extent of planning and coordination, as discussed later.

In general it can be concluded that international funding mechanisms have had a relatively small impact on oceanographic research per se, which continues to be funded primarily at the national level. The principal importance of international funds has been in assisting developing countries to acquire technical competence in the development of their marine resources.

Planning and coordination mechanisms

Public international councils may exist for coordinating the work of other intergovernmental organizations. Until recently, there has been only one such council in the field of ocean affairs, the Sub-Committee on Marine Science and its Applications of the Administrative Committee on Coordination (ACC), Economic and Social Council of the UN. The ACC, composed of the administrative heads of the various UN bodies, has the functioning of coordinating activities within the UN system, and the Sub-Committee has provided the means for consultation at the working level among representatives of the various agencies. It has not been a very effective council since agency representation is below the policy-making level and, because of different budgetary periods and practices among the agencies, coordination has been limited to the exchange of information on plans and programs.

New, and potentially more effective, mechanisms are now being developed through the IOC. These result from the action of the UN General Assembly in approving a long-term and expanded program of oceanographic research (the Expanded Program) and in asking IOC to intensify its activities with regard to coordinating the scientific aspects of this program. Pressure has also developed from other UN bodies in addition to UNESCO for a stronger role in supporting IOC. This has led to establishment of an Intersecretariat Committee on Scientific Programs Relating to Oceanography. Thus the evolution of IOC now underway should result in a more effective means of coordinating the scientific ocean activities of the interested UN organizations.

As yet there exists no mechanism for coordinating UN activities with those of intergovernmental bodies outside the UN system (such as ICES, IHO and ICNAF). Because a given country may belong to a number of organizations, it might be expected that some coordination would take place at the national level, that is that the country would implement different aspects of its single national ocean policy through the different organizations. But this does not usually happen, perhaps because national links with the various organizations are often diverse and poorly coordinated. The problem becomes critical with development of the Expanded Program which should encompass international cooperative programs of all sorts, not just those developed within the UN system.

Private international councils may contribute to the planning and coordination of intergovernmental programs by monitoring and evaluating such programs and by presenting their views to IOC and analogous bodies. In addition, the private councils may organize limited cooperative programs of their own. Such programs may consist, for example, of field trials for the intercomparison of equipment and methods, or joint studies by several laboratories of a specific region or problem. Large-scale investigations, as noted earlier, appear to be more effectively arranged through intergovernmental agreements.

There are several councils with oceanic interests within the International Council of Scientific Unions. Among these, SCOR is the most important. The Scientific Committee on Antarctic Research (SCAR) has a subsidiary group concerned with Antarctic oceanography. The Special Committee on the International Biological Program (SCIBP) has a section on marine productivity. There are also inter-union bodies such as the Upper Mantle Committee or the Commission on Geodynamics which may develop scientific programs with a significant ocean component. In general, these bodies work closely with SCOR on oceanographic matters.

Communication mechanisms

Scientific organizations were first created to improve communication among scientists, through the convening of scientific meetings and the publication of scientific papers. In most countries where ocean research is conducted there exist such societies, either specifically intended for oceanographers or, more commonly, as sections of organizations concerned with more basic disciplines. In the countries most active in ocean research, there are innumerable opportunities to discuss and publish scientific results.

Internationally, ocean-oriented societies are, for the most part, components of the International Council of Scientific Unions, the principal ones being the International Association for the Physical Sciences of the Ocean (IAPSO) of the International Union of Geodesy and Geophysics, the International Association of Biological Oceanography (IABO) of the International Union of Biological Sciences, and the Commission on Marine Geology (CMG) of the International Union of Geological Sciences. Other societies with different or broader interest may also at times consider oceanographic topics. Examples include the International Association of Seismology and Physics of the Earth's Interior, the International Association of Meteorology and Atmospheric Physics and the International Association of Volcanology and Chemistry of the Earth's Interior of the International Union of Geodesy and Geophysics, and the International Association of Geochemistry and Cosmochemistry of the International Union of Geological Sciences.

SCIENCE AND OCEAN POLICY

To recapitulate, marine scientists are interested in the development of international ocean affairs as it affects the rate at which scientific knowledge can be gained and as it affects the benefits to mankind from the ocean. The future effective use of the ocean and its resources appears to be increasingly dependent on understanding of ocean processes and phenomena. A number of ways in which international action can further oceanographic research have been enumerated, and the existing mechanisms for achieving such international action have been examined. In view of these considerations, what improvements in intergovernmental ocean policies are desirable?

To acquire scientific knowledge at a satisfactory rate, it is first necessary to strengthen national capabilities to do oceanographic research. Oceanographic capability is measured by the availability of competent, trained scientists able to devote their attention to research and by the support given them in terms of funds, facilities and assistance. Thus there must be means for training such scientists, permanent posts and assured careers in marine science, and adequate budgets for ocean research. Achievement of these conditions assumes a national commitment to such research. Suitable national mechanisms for planning and coordination are also essential.

Although there is some experience in technical assistance intended to improve the oceanographic capability of the less developed nations, attempts to date have not been notably successful. The relative failures can be attributed as much to lack of significant commitment and investment on the part of the recipient country as to inadequacies in the assistance provided. But much still needs to be learned about the most effective ways to transfer oceanographic skills and knowledge, and the careful review and evaluation of experience in this field would greatly benefit future programs of assistance.

We have seen that international programs can usefully pool national efforts and interests in oceanographic research as well as in application of such research to the conservation and management of ocean resources. Before discussing how the mechanisms for joint action can be improved, it is important to comment on the national links with such programs. In a field so complex as ocean affairs, it does not seem likely that there will ever be a single monolithic organization capable of dealing with all intergovernmental aspects of the problem. Rather, there will continue to be an assortment of specialized mechanisms, hopefully more successfully coordinated than at present, but still presenting a variety of receptors for national contact.

Cooperative ocean investigations can be organized at different levels, from informal agreements among scientists to formal worldwide intergovernmental arrangements. In this scale of complexity, intergovernmental mechanisms first appear at the regional level. Where regional arrangements are made among countries with adequate oceanographic capabilities, as in the north Atlantic, concerted action can be very effective. But most countries bordering the sea do not have an adequate oceanographic capability, and the collective study of vast areas of the world ocean requires for its support an organizational structure of global scope.

Recognition of this need led to the establishment within UNESCO of the Intergovernmental Oceanographic Commission, the only worldwide intergovernmental organization wholly devoted to scientific investigation of the ocean and its resources. The IOC has functioned only since 1961 and, considering both its youth and the financial constraints under which it has operated, has achieved remarkable success. At present IOC is undergoing a transformation involving a deeper involvement of other UN agencies in addition to UNESCO and the acceptance of broader responsibilities on behalf of the UN system of organizations. It remains to be seen whether this desirable broadening of support will provide the IOC with the necessary strength to discharge adequately its functions.

In the long run, this seems unlikely. It is true that present funds could be used more effectively if pertinent UN agency programs were more closely integrated with those of IOC. This should permit elimination of overlap in staff activities, mutual reinforcement of strengths in staff and programs, and combination of certain program elements such as fellowships, training and publication. But the total funds available in the several agencies concerned are entirely inadequate for the work to be done. Furthermore, the oceanographic components of the budgets of existing UN agencies cannot be expected to grow rapidly since this growth can only be at the expense of other programs which also have their active constituencies.

Therefore, a need exists for a separate intergovernmental agency to deal with the scientific and engineering aspects of ocean affairs. Such an agency could provide a much-needed focus on ocean research and development by bringing together the various international interests in these problems, and could provide the leadership required to achieve a new level of funding. The present steps to strengthen IOC should be seen as necessary preliminaries to the splitting off of a separate organization.

Proposals have been made for intergovernmental ocean agencies intended to deal with all aspects of international ocean affairs, from political and legal problems through the management and conservation of marine resources to the scientific investigation of these resources and of the ocean itself. Unfortunately, ocean science would probably suffer in such a monolithic organization. Experience in the IOC already has shown that consideration of legal and associated political problems tends to dominate the sessions and to influence the membership of national delegations by diluting out their scientific interest and competence. In order effectively to "promote scientific investigation with a view to learning more about the nature and resources of the oceans," it is necessary to have a specialized organization devoted to this task and capable of attracting the enthusiastic interest and support of qualified marine scientists throughout the world. Such an organization, adequately funded and staffed, should serve as the technical counterpart of whatever management body may ultimately be established.

A problem that must be solved by IOC or its successor is that of providing coordinating links between international programs within and those without the UN system. Oceanographic efforts available for cooperative investigations will continue to be pooled in regional combinations, while the results of these investigations will accrue to the benefit of all interested parties. Thus the regional programs are integral parts of the over-all study of the ocean. A worldwide ocean science organization, such as IOC, must recognize the value of such programs however they may be organized and must find ways to support them. The present IOC cooperation with ICES in the north Atlantic and with several organizations in the Mediterranean should provide useful experience in developing the necessary coordinating links.

Because of the unity of the world ocean and the processes operating therein, many scientific studies cannot be restricted to locally confined regions. The need to look at the ocean as a whole is accompanied by the need to be able to conduct research in any part of the ocean. The freedom of scientific research on the high seas has not been seriously restricted until recently, with activation of the Convention on the Continental Shelf. Even now, only isolated difficulties have arisen, but the potential for serious limitations is provided in several paragraphs of this convention as they may be interpreted by suspicious coastal countries. The opposition of many scientists to proposed international regimes for the ocean is based on the danger that these regimes may deliberately or inadvertently result in restrictions on the freedom of research.

It should be the goal of all maritime countries and of their intergovernmental associations to ensure that scientific exploration and research can be freely conducted everywhere in the ocean beyond the limits of internal waters. Exploration is a fundamental part of research, and attempts to distinguish between exploration and research based on the motivation of the investigator are meaningless. Where the waters or sea floor are subject to national jurisdiction, it is only proper that this freedom be accompanied by a responsibility to afford the coastal country the possibility of full participation in the research and full access to its results. On the high seas, although these conditions are not imposed, the tradition of scientific research has always favored the full exchange of data and of their interpretation.

It is remarkable how little attention has been paid in international discussions of ocean affairs and of ocean regimes to existing oceanographic information and activities, to the need for, and means of, acquiring additional necessary information, and to the interest of marine scientists in these matters. Oceanographers are exploring ways whereby their views can be more effectively formulated and made known. This effort is based on a re-evaluation of the relationships among the societies and councils within the International Council of Scientific Unions. Similar efforts should be made by intergovernmental bodies to reinforce their communications with the ocean science community. A number of elements enter into international ocean affairs, of which ocean science and research must be counted among the most fundamental. Marine scientists should be mobilized, together with their colleagues in the engineering and social disciplines, to work closely with governments on these problems of universal interest and importance.

ALEXANDER KING

6 International Cooperation in Science and Technology

It is not surprising that a considerable activity of international co-operation has been generated in the field of the natural sciences and technology. Contemporary scientific development in the individual country or individual laboratory is part of an international mosaic. Each research worker draws upon the work of many colleagues throughout the world for concepts, experimental methods, instrumentation, and data. His own work is published freely in journals which circulate throughout the world's system and forms in turn an element of input in the work of others. The international system of scientific communication is very rapid and efficient; the world output of original work is scanned, abstracted, electronically indexed, and made available so that substantially no new contribution is lost or its significance ignored for long. The system's chief weakness is in fact that of overloading; there are so many new papers and so little analysis of quality that selection of the significant from the trivial is difficult and it is at any rate initially assessed mainly on the international reputation of its author in the at times exceedingly small circles of specialists, frequently in acute competition.

The principle of free international publication of results in basic research operates universally even in times of acute nationalism.

ALEXANDER KING is the Director-General for Scientific Affairs, Organisation for Economic Cooperation and Development, in Paris, and an Associate Fellow of the Center for the Study of Democratic Institutions. Mr. King has asked us to state that the views expressed in this chapter are his and not necessarily those of the O.E.C.D.

As with all scholarship it preserves the illusion of purity—it is undertaken for motives of intellectual curiosity and not in the hope of gain from its application. The illusion is however less and less easy to maintain. Ever since World War II, governments and industry have recognized the value of new scientific knowledge and have financed it in the hope of national profit in the attainment of their goals, mainly those of defense, economic growth, and national prestige, so that gradually the government in most countries has become overwhelmingly the most important patron of research. However, basic research is not a very clear-cut or easily evaluated national investment. Because its results are internationally available, they are internationally exploitable and hence the real national investment is the capacity to exploit the world product of new research and not just that originating in the country in question. In smaller, industrialized countries such as some of those of western Europe only some five percent to ten percent of the research exploited for economic purposes originates domestically. Yet it is dangerous for a country to attempt to exploit the research of others without a commensurate contribution to the world's scientific output. Only through indulging to a substantial extent in research will a country possess a level of scientific awareness which enables it to recognize and select from the world product that small proportion which is significant for its own needs. Investment in research therefore yields only indirect dividends—the direct advantages may be gained in far distant countries and at much later periods. It is only recently that the place of research in the total process of innovation has been recognized and still many decision-makers and many scientists assume that more research means rapid economic growth. This may or may not be true, and even when it is exploitation success depends on many other factors, including entrepreneurial ability, capacity for technological development, availability of risk capital, suitable fiscal arrangements, levels of education, and even national psychology.

Formerly research was classified as basic or applied, the former often being assumed to be pure research. This distinction is less and less true. Especially in advanced industrial firms or mission-oriented government agencies, working near the frontiers of knowledge, new data or new scientific theories are required to extend the frontiers for applicational purposes. An ever greater proportion of basic investigation is therefore oriented research, oriented, that is, towards practical use. The contrary is also true; advanced fields of basic inquiry such as nuclear research, high-energy physics, radio-astronomy, and space research aimed at the extension of knowledge for its own sake—the so-called big sciences —require complex and costly installations and instrumentation for the purpose which demand the creation of new technologies to permit "pure" research. Furthermore the instruments and methods of basic and applied research in advanced fields are similar so that a research worker may on one occasion be undertaking basic work and on another functioning as a scientific engineer.

INTERNATIONAL COOPERATION IN RESEARCH

While essentially part of an international system, most of the research done in the world is by individuals or groups working in universities, firms or institutes which are nationally based, although the emergence of the multinational firm, often with strong research activities, is somewhat altering the picture.

There is however an increasing number of organized activities at the international level, either arranged by the scientific societies or organized on an intergovernmental basis. Since World War II organizations for international research have proliferated very rapidly and frequently not within the framework of any discernible policy of international cooperation.

There are three main justifications for international research cooperation which do not necessarily apply to all existing cases.

These are:

1. Where the very nature of the research topic is such that the field of inquiry cannot be restricted to a single unit of national sovereignty;
2. To make possible the cross-fertilization of ideas, particularly in newer and specialized fields where only an internationally assembled team is likely to provide the necessary critical mass of activity (cerebral or practical);
3. To make possible, by cost sharing between countries, the

provision of the very expensive equipment and operation required by "big" science.

In addition there is at times the idealistic motivation that by bringing together people from different countries to work together world peace is encouraged.

The earliest of the great international projects for scientific cooperation were of the first of these three categories. In 1824 the astronomer Bessel asked several laboratories in different countries to combine their efforts in the compilation of a sky chart. This resulted in a scheme under the auspices of the Berlin Academy, whereby a large number of astronomers from a variety of European countries took responsibility for charting and describing a specific area of the sky. This scheme continued and extended with the application of photography to astronomical measurement, and in 1887, a permanent international commission was set up, which took more than fifty years to complete its work. This was a purely scientific endeavor which involved no intervention from governments.

The organization of "international years" for specific programs of research was more complicated in that they involved several disciplines and governmental assistance. In the first of these, the International Polar Year of 1882–83, eleven national expeditions took part as well as the observations of thirty-five countries. This was the forerunner of other international years such as the International Geophysical Year of 1957–58 and that of the Quiet Sun of 1964–65.

Until the nineteenth century science was seldom hampered by frontiers—even in times of war—as instanced by the famous cases of Humphrey Davies who travelled to Paris to give a lecture at the Institut de France during the Napoleonic war and of Benjamin Franklin who granted safe conduct to Captain Cook's vessel during a period of strained relations between Great Britain and the United States. However from the nineteenth century, national frontiers and languages, despite the universal and open character of the scientific community, began to constitute a barrier to scientific mobility and cooperation.

Expenditures on research have increased exponentially throughout the last two-hundred years and it is only now that they are flattening out, while in a few countries such as Japan and Germany the steep increase still persists. The greater part of the research activity has, therefore, been undertaken since the second world war and financed by governments in support of goals of defense, economic growth, and prestige. Governments having become the piper, naturally call the tune, however discreetly this may be done, and hence the relevance in their eyes of scientific programs and the articulation of research with national policy in general have become major preoccupations. This has two consequences for international research cooperation. First, *cooperation* is encouraged and even demanded for reasons of resource scarcity. Except in the two biggest countries, men and money for research are greatly inferior to the promise which new scientific and technological knowledge seems to hold for the national advantage, and even in the USA and USSR this is gradually becoming the case. For smaller nations with advanced industrialization, such as those of western Europe, it is only by the cost-sharing device of international schemes that they can hope to keep abreast with world developments in expensive scientific fields such as high-energy physics, radio-astronomy or space. At the same time they feel that unless they are engaged in such advanced research activities, they will gradually fall back in industrial development. The second factor at first sight appears to operate in the opposite sense. The defense, growth, and prestige objectives of nations are traditional fields of international *competition,* hence the main government-financed elements of national research are mounted for reasons of international competition. In contemporary world science then, cooperation and competition march hand in hand and interact in the increasingly political nature of the research process.

INTERNATIONAL ORGANIZATIONS

International organizations for research are of two types, those organized by the scientists themselves or their institutions and the intergovernmental organizations. For the reasons mentioned above, the latter has become the most important although not necessarily the most effective category.

The main nongovernmental organization on the international plane is the International Council of Scientific Unions (ICSU), a federal body which unites a number of scientific unions for particular scientific fields such as chemistry, physics, astronomy, and geodesy. This system, which has now a long history, evolved from early association between the various national academies. The nongovernmental organizations generally do not undertake actual international research programs as such, but they organize meetings and symposia, facilitate the movement of scientists, the communication of results, and the standardization of units and nomenclature. ICSU has however been the initiator of many international research schemes including the international years for particular topics. Such organizations enjoy the advantage of great freedom in the choice of their programs and in attracting the interest of highly qualified personnel. Furthermore their flexibility of structure enables them to respond to new needs arising from scientific change, which they are in a position to identify sooner than intergovernmental bodies. At times they, or individual members, have played an important part in persuading governments to initiate international schemes and in drawing up the initial structures and programs for intergovernmental organizations. The main problem of these flexible bodies is of attracting sufficient financial support from governments, although general subventions from UNESCO have helped greatly. There are in addition a few international scientific stations, such as the Jungfraujoch Scientific Station founded in 1930, which have useful and original activities.

Intergovernmental scientific organizations, although they have mushroomed since 1945, are no new concept. The International Bureau of Weights and Measures at Sèvres was founded in 1875 as the custodian of the meter and other metric standards. This organization maintained a research laboratory from the outset and has kept up well with the developments of science, including important work on nuclear standards.

The period immediately after the war saw a proliferation of international scientific activity. New organizations sprang up on both official and unofficial bases, with little relationship between them and in the absence of any coherent policy. The motivation varied greatly; much of it came from the scientists themselves, who were anxious to quicken the development of their own special fields or to obtain by international cost-sharing access to expensive equipment which otherwise would have been unobtainable. There was little clear statesmanship of science, either from individuals or from their academies, rather a bevy of separate, enthusiastic but unconnected vested interests.

Between 1945 and 1955, fifty-eight new international scientific organizations were created. In 1963 there were more than two-hundred and fifty nongovernmental and fifty intergovernmental organizations directly engaged in scientific activities. These figures can only be approximate as no precise inventory is possible in the absence of an agreed definition. International organizations that may properly be described as scientific include both those which carry out scientific programs or coordinate research activities as well as those which discuss scientific topics and organize conferences but undertake no original research. Each year several thousand international conferences and seminars on scientific topics are held.

Soon after the creation of the United Nations, its Social and Economic Council put forward ambitious plans for the establishment of international laboratories in a number of fields, including study of the brain and a computation center. Little came of this initiative as its real justification was of international idealism and the individual projects had no clear cost-sharing or intellectual basis which could fit with the plans of individual governments. The specialized agencies of the UN however soon began to concern themselves with the fostering of scientific research within their respective fields—this has been particularly marked in UNESCO (1945), the World Health Organization (1946), the World Meteorological Organization (1947), and the International Atomic Energy Agency (1956). All these bodies are, of course, worldwide in character and have a membership of upwards of a hundred nations, the great majority of which are underdeveloped countries. International cooperation between such large numbers of countries is exceedingly difficult and also expensive to achieve in a realistic sense, the inevitable compromise within the diversity of environments, levels of development, attitudes, needs, and interests all too easily proving to be bloodless. The numerical dominance of underdeveloped members likewise pushes such organizations towards aid rather than cooperation between equals. This is of course an important function which has greatly assisted the transfer of science and technology, but it does not assist in maintaining a level of sophistication of discussion and exchange which the relatively small proportion of nations which produce the greater proportion of new research require for their own purposes or for the advancement of science as such.

Of the specialized agencies, UNESCO is the one with the broadest function and hence has taken the most comprehensive approach to scientific cooperation. Some sixteen percent of this organization's ordinary budget apart from technical aid is devoted to such work. This includes subventions to the international unions already mentioned, coordination of international programs in the life sciences, oceanography, etc., and in research on natural resources (that is, study of prospecting methods, studies of arid zones, humid tropical zones, etc.). More recently, in addition, UNESCO has concerned itself strongly with questions of international science policy in relation to the economic growth and social advancement of countries in the course of development.

THE APPROACH OF THE ORGANIZATION FOR ECONOMIC COOPERATION AND DEVELOPMENT (OECD)— AN EXPERIMENT IN COOPERATION

At first sight it may appear surprising that an international organization, dominantly economic in its interest, should support important programs in science and education. The fact is that OECD is concerned with these subjects basically in relation to economic growth and social development as important items of national investment which, although of long-term effect, contribute greatly to the vitality of society and the economy and have a significant role in the effective utilization of the many resource inputs of the economy, namely capital and labor. The OECD membership moreover is the dominant producer of new science and technology, providing nearly seventy-five percent of the world's scientific effort.

OECD activities in science and education therefore attempt to link these subjects with national social and economic objectives and to develop them in a progressive policy framework. The programs of OECD therefore have a flavor very different from those of most other organizations in that they are much less concerned with international activity as a good thing in itself; but rather the OECD uses international experience, confrontation, emulation, experiment, and demonstration to present the twenty-two member countries with new thinking and ingredients of policy change which they can, if they feel it appropriate, incorporate within their own traditions and institutional patterns.

This being so, the programs aim at being promotional and catalytic, at identifying new concepts, new institutional approaches, and new methods, surveying, comparing, and developing these, undertaking pilot and demonstration projects aimed at encouraging innovation in the member countries. The approach is that of innovation; the Science Directorate of OECD attempts to terminate each element of work once a sufficient impact has been made. Should individual items be carried to the extent that they threaten to become permanent service features, there would be a gradual consolidation of programs along lines which have proved successful, so that within the limits of more or less fixed resources there would be little effort available for new promotional probes. Thus a few years ago an incisive reappraisal of the programs resulted in about one-third of its items being cut, including topics such as the reform of science teaching which had proved very successful.

Such abandonment of successful experiments and a continual scanning of new opportunities is difficult and painful. It is hard to drop one's successes and continue with the speculative and only partly known. Furthermore it necessitates a high degree of staff mobility because different skills are needed from year to year at a considerable depth of specialization. This is why the directorate works to such a large extent with the help of temporary consultants; these may be professors on sabbatical leave, experts of world reputation who work for a few months or intermittently. In this way a flexibility and expertise is assembled which could never be provided by a fixed international staff alone.

Staff and consultants come from many intellectual disciplines, from the natural sciences, the social sciences and technology. Because the work is problem oriented, this multidisciplinary and interdisciplinary attack seldom gives rise to problems of demarcation or intellectual sectarianism. Such multidisciplinary approaches are very rare in Europe.

SCIENCE POLICY

Although science and scientists played a dominant part in the war and both governments and industrialists were convinced of the importance of a science policy for peacetime economy, the enormous extension of research activity in the years 1945 to 1960 in most industrialized countries was a spontaneous burgeoning with little real policy framework. Science was regarded as a good thing; those countries with a massive research activity were assumed to have a high potential for economic growth. Scientific expansion had in fact flourished on a more or less intuitive conviction of its importance, with a heavy mystique greatly encouraged by the scientists themselves.

By about 1960 expansion had proceeded to the extent that national expenditures on research began to demand a considerable fraction of national incomes, and hence it could not long escape the financial and policy scrutiny which is inevitably given to all main government expenditures.

OECD has taken a pioneering role in attempting to develop concepts of science policy, in stimulating the creation of national political structures appropriate to the elaboration of healthy national policies for science and to assist in the demystification of the subject by careful analysis of the elements of policy and particularly the relationship between science and the various national goals.

In 1961 the Secretary-General of OECD set up a high-level group of scientists and economists to discuss the concepts and mechanisms of science policy. This group, in its report "Science and the Policy of Governments" pointed out that there are two aspects of the problem. First, the need to establish policies for the management of science and the allocation of resources for its support between the many alternative claims. Second, the need to consider the impact of science on other elements of national policy.

The Organization, in following up the report, has concentrated mainly on policies for the management of science and especially on problems of resource allocation because the extent of research activity desired by most countries (and certainly most scientists) is much greater than the resources which can be made available. With regard to policies through science stress has been mainly on the relationship between science and economic policy.

The report proposed two main recommendations. First, that each country should establish a mechanism for the elaboration of its national research policy, and second, that OECD should convene a meeting of the ministers concerned with science and its organization to discuss and deepen understanding of what science policy is and could be.

These recommendations have been largely followed. The first ministerial meeting was held in 1963. Only a few countries had at that time already appointed Science Ministers, so that about two-thirds sent their Ministers of Education. The chairman of this first assembly of science ministers was the Belgian Prime Minister who was also Minister for Science. By the time the second ministerial meeting took place in January, 1966, about two-thirds of the countries had made some special arrangements for science policy and had ministers responsible. At the third ministerial meeting held in 1968, most countries had ministers specifically designated for science policy, while several of these were accompanied by their economic colleagues.

The ministerial discussions included the nature of science policy, how resources for science are allocated, the place of fundamental research and of the social sciences respectively, on the policy of governments, the relation between research and economic growth, the function of governments in industrial innovation, international cooperation in research and problems of the transfer of technology, etc.

As a basis for considering resource allocation, quantitative data are essential as in other fields. OECD has, therefore, given considerable attention to the collection of statistics of national expenditure on research and development (R & D) both in terms of finance and of professional manpower. The data published by various countries led to some misleading comparisons because of different definitions of the various categories of research and different methods of collection and analysis of the data. As a first step, therefore, a group of experts from the OECD countries was asked to formulate a series of definitions which might be generally acceptable and to search for a uniform method of census. This was done on the basis of a first draft by a consultant and led to the compilation of the so-called Frascati Manual which after discussion by the statisticians of each country was modified and finally adopted by the OECD Council as a basis for country action. This made possible the acceptance of an International Statistical Year for Research and Development which has provided for the first time comparative data, broken down by sector, distinguishing between university, government, and industrial research. The Frascati Manual has now been revised in the light of experience gained and will eventually be extended to include social science research.

RESEARCH COOPERATION

An advanced and successful example of an international system is the European Nuclear Energy Agency (ENEA), a subsidiary body of the OECD. The board of ENEA is essentially a forum for the exchange of views on priorities and for negotiating the shape and finance of specific schemes, termed joint undertakings. Some of these, such as the Dragon and Halden (Norwegian) reactor programs, for each of which the particular country in which the reactor is situated assumes its management with financial and scientific participation from a group of interested countries. One undertaking, EUROCHEMIC, is in reality an international company for the treatment of used fuel elements.

In many fields of pure and applied science, centralized institutions are not necessary on financial grounds. OECD has therefore encouraged the shaping of programs agreed in common by groups of countries, and with elements of the program carried out in universities and other national institutions. A considerable number of limited, practical schemes have been carried out in this way during the last ten years under OECD initiative; subjects include marine corrosion, metal fatigue, cutting and shaping of metals, biological deterioration of materials, air and water pollution and research on road construction and road safety.

In a number of other instances independent bodies have been created by OECD for particular tasks. For example, after a series of seminars which took place over several years on the management of research programs, industry was invited to create its own organization for the purpose, and a European Industrial Research Management Association (EIRMA) was created, to which some seventy of the largest European firms now adhere. This body will function in close relationship with the Industrial Research Institute in the United States.

A further example of the creation of independent or semi-independent institutions arising from these policies is that of the Centre for Educational Research and Innovation (CERI) which is at present financed by subventions from foundations and industry. Plans are also nearing completion for the creation of an International Institute for the Management of Technology.

The direction of OECD work on research cooperation has changed of late. Instead of stimulating a large number of small individual cooperative schemes as described above—and some forty separate groups have been involved—the committee is concentrating its effort on a smaller number of larger fields where a sustained interest and effort by governments is required in which science and technology have much to contribute. These include urbanism, advanced transportation technologies, the management of water research, preservation of the environment, and the development of new materials. These represent national activities of growing importance and great complexity, subjects on which the scientist, the engineer, the economist, the sociologist, and others have contributions to make but where the problems are immune to attack by a single discipline alone. In most cases responsibility for such problems does not fall clearly into the competence of a single government department and it is often difficult to create comprehensive policies.

NEW PERSPECTIVES

During the decade of preoccupation with science policy the situation has changed radically. During the 1960's for example the gross national product (GNP) of the OECD region has increased by fifty percent, and a still greater economic growth is envisaged for the 1970's. At the same time, both as a consequence of growth and in relation to the technological development which made it possible, the impact on society and the individual has been profound and by no means uniformly positive. Not only is there the shadow of biological warfare and of the possibility of the manipulation of mankind by the new discoveries of the biological sciences, but there is an increasing feeling amongst a large proportion of the population of a deterioration of the quality of life. The sterility of individual life in great cities and their colorless suburbs, deterioration of the environment, frustration with regard to commuting and transportation difficulties in general, the distance and inhumanity of central decision-making—these and many other factors are leading to an increasing extent of alienation and rejection of authority, especially in a period of vanishing faith in religion and lack of confidence in the democratic process.

Without taking an apocalyptical attitude, the question must be raised as to whether our societies can sustain a further period of rapid economic growth without breakdown, unless quite radical new measures are taken and new thinking accepted. Voices are already being raised against continuing the stimulation of economic growth, just as they are for a moratorium on research activity. Yet this is no solution, the problems are much more fundamental than those of growth which is still required if we are to meet the needs of society, for example, with regard to education and urban renewal, to say nothing of the pressing problems of the underdeveloped world. Already there is a recognition on the part of many governments in the industrialized countries that economic growth should no longer be regarded as an end in itself but as a mechanism to provide the resources necessary to achieve a whole range of policies of national development.

For science itself major readjustments are called for. Its very nature as well as its objectives have altered considerably during the decade in question. Some of the main elements of change are the following:

1. Just as primitive ideas of the influence of new scientific research are being replaced by a more sophisticated, although still inexact concept of the process of innovation, so the impact of science on other sectors of policy appears to be a more complex and subtle mechanism than hitherto assumed. In medicine, agriculture, urban development, education, communication and travel, new discoveries uncover new possibilities, but also distort the balance of the system in question, scattering unexpected and often unwanted side effects, producing new colonies of problems. In these and other fields the central problem is that of assimilating the new knowledge with the complex, interacting systems—once again the little understood mechanism of innovation, whether technological, educational, medical, or social.

2. Problems of multidisciplinarity are growing. New borderline fields arise between the traditional sciences, they themselves develop interfaces and new orientations. Science is no longer to be regarded as a series of discrete although related fields, but a dynamic system, ever in transition. Such a situation is particularly difficult for the Europeans to cope with, owing to rigid faculty systems in the universities, which, in spite of all the pressures, seem impervious to change. On the other hand the complex problems to which the scientific approach is attempting to find solutions, necessitate multidisciplinary and simultaneous attack from the natural, economic and behavioral sciences. If multidisciplinarity is difficult between the natural sciences and engineering, how much more remote is the hope of a genuinely combined approach with the social sciences.

3. Science with its exponential growth sustained over two-hundred years is an *enfant gâté* amongst the fields of learning. Now that the growth is flattening off all sorts of uncomfortable readjustments are required, as the present situation in the United States indicates. In the present situation questions of priorities and national goals are of paramount interest, but many new problems will begin to arise. For example with the constant expansion of scientific resources, the average age of research workers has

always been very low; this will no longer be the case, with consequent implications for creativity and retraining.

4. The public image of science has dimmed. Its mystique has largely evaporated and its negative manifestations exaggerated, often to a point of hysteria. The formerly desirable objectives of national affluence have been relatively and unevenly attained, but the dirty face of affluence has been exposed to exhibit islands of poverty and despair, prejudice, underprivilege, the monstrosity of urban life and the debasement of individuality. There is also the uneasiness that science and its methods are creating a technocratic elite, powerfully bent on soulless planning, destroying or ignoring cherished (although often outmoded) values and almost completely lacking contact with the mass of ordinary citizens, that is, with both the radicals and the "silent majority." However exaggerated these symptoms may seem they cannot be dismissed, and science will have to strive hard to re-establish its reputation for objectivity and create a collective statesmanship rather than to continue as a clutch of vested interests.

5. Finally there is growing understanding of the dangerous gap between the rich and the poor countries, of the fact that research seems mainly to make the rich richer, while at the same time the interacting problems of world food, population, and health are susceptible to scientific attack. This is persuading many scientists and especially the young that there is something seriously wrong with the research priorities of the so-called advanced countries.

The nature of the problems of contemporary society is so complex and compounded of political, economic, social, and technological elements to say nothing of the values questions they raise, that there is an urgent need to look at the world system as a whole and to strive to understand something of the interaction of their constituent parts. Of course these matters are in their totality far beyond the range of science and technology but to some extent they result from the present technological upsurge: they all have important technical elements and they are accessible to research attack and to analyses by the scientific method in formulating and delineating their nature and the importance of their interactions.

These areas of concern have three features in common. First, they have a global character and appear to arise at a certain stage of development irrespective of the political or economic system. Second, they are multivariant in their nature and in particular have economic, social, and technological elements which demand multidisciplinary attack. Finally, they are interconnected and proposed solutions to one problem may give rise to unexpected side effects elsewhere. We are reaching a stage in the affairs of society where there are no longer well-defined and isolated problems, nor can we expect clear-cut solutions. To persist in a problem-by-problem approach is at best likely to remove a few symptoms without discovering the basic cause and at the worst by curing one symptom to give rise to many more appearing unexpectedly in quite other parts of the system without the connection being understood or even recognized.

One of the initial tasks in the use of science for development would be its assistance in the formulation of the goals of society, not in a decision-making role but through formulation and analysis. Here again the total systems approach becomes important, since at present it is in the attempt separately to attain a series of national goals that the so-called side effects of technology become most obvious. These may be either positive or negative; on the one hand, the spin-off both in systems-management competence and in hardware for defense and space programs and, on the other, the undesirable effects of technological development. The integration and co-management of goal attainment is of course extremely difficult in existing governmental systems with their vertical sector structures devised for a simpler age of compact problems and solutions. Examples of inter-goal conflict are numerous, and the hidden or unidentified cases must be still more frequent.

Another matter on which action is urgently required is the matter of assessment of the probable consequences of a major technological development before it is effectively launched. This is a somewhat lengthy and costly matter, but techniques for its accomplishment already exist. The related problem of forecasting is also in a stage of quick development, not merely technological forecasting in the simple sense as it has been evolved by defense groups and now applied in the industrial context, but social forecasting and integrative forecasting which combines the various approaches now beginning to be attempted in isolation. This wide subject of scanning the future prospects is not yet accepted as a general governmental function, but its necessity is now strongly argued. As so many of these matters are of global importance and as the procedures are heavy and costly, there is much to be said for an international approach to such matters in order to spread the costs and especially the time of those possessing the very rare skills involved.

Much greater importance will have to be given to the general planning function, to the concept of inventing the future, of choosing between alternative futures and attempts by normative feedback methods to planning in a dynamic sense, adjusting goals as new elements of information and experience accumulate and as new scientific possibilities emerge. There is a strong appreciation amongst those concerned with the development of new planning approaches to shed as completely as possible the technocratic image. The Bellagio Declaration, issued by the participants of the OECD seminar on planning and forecasting makes this clear. It accepts that in planning, appreciation of human values must transcend technocratic objectives, it makes clear the need for "participative" planning and implementation and the necessity for developing new structures. It notes, *en passant,* that a scientific approach in planning today often serves merely "to make situations which are inherently bad, more effectively bad."

On the national level, this integrative approach to science and society will be very difficult to achieve. The existing science ministers and their consultative apparatus are hardly appropriate as a departure point. There is a need for new structures and a central decision point which would enable comprehensive and coherent attention to be given to the longer-term problems which have an all too obviously shrinking time scale. Economic policy, social strategies, educational development, and science policy—these can hardly be pursued separately for much longer, but dynamic structures of government and even the concepts of such are hardly, as yet, subjects of discussion.

SUMMARY

We have briefly scanned and analyzed the experience gained in organizing international cooperation, both intergovernmental and between research scientists and their societies. Such cooperation, increasingly required to extend national resources for R&D through international cost-sharing, has developed rapidly during the past twenty years—especially in Europe—but has proliferated mainly in the absence of a policy framework.

The scientific work of OECD has been taken as a case history of the development of science policy and the promotion of research cooperation. In particular, the need for change in science policies is emphasized to enable it to face the multivariant and complex problems now facing society, including those of the environment. The need for a multidisciplinary and sustained attack on these problems is stressed, and especially the need for governments to create political structures which will enable medium-term strategies for economic development, education, science and technology, and social policy to be integrated.

PART TWO:

DEVELOPING THE SEA FRONTIER

Experienced underwater engineers paint a fascinating picture of the oil industry as a vast, land-oriented industry, advised by land engineers leaping into the sea as if the seabed were a flat prairie where skies are always blue and the seaweed as high as the sea elephant's eye. Many of the supposedly unexpected disasters were preventable in the considered view of oceanographers familiar with the medium, whose advice went unsought or unheeded. To quote a spokesman for a major oil company, "the oilman pretends the sea is not there."

It is not only the oilman. It is the international lawyer, the disarmament expert, the diplomat, the statesman as well. It is a curious paradox to pretend the sea is not there while trying to build an ocean regime. This contradiction, explicit or implicit, is a deep-seated impediment to progress in establishing an international ocean regime.

The drawing of boundaries, the carving up of the ocean floor into blocs to be signed over to control by nations, the notion of a simple "registry of claims" bureau are examples illustrating the "terrestrial" approach to ocean problems.

Under this approach, the structure of the regime is supposed to be determined by two perimeters: its territorial boundaries and the amount of resources available beyond these boundaries in the area beyond the limits of national jurisdiction.

Yet the boundaries, projected on and through the fluid medium, waver, and not all the wisdom stored in all the computers in the world can fathom the wealth of the oceans. Estimates, both of actually available resources and of their potential development, differ widely; and the gaps in our knowledge and the failure to integrate fragmented bits of information are deep and dark.

Dr. Schaefer's, Dr. Kasahara's, and Dr. Laque's chapters present descriptions of the present situation and its linear projection into an immediate future. They raise many questions. They pose many paradoxes.

Boundaries and quantities of resources are static concepts, to be mapped on classical earth maps. The marine maps Dr. Wenk postulates in his chapter do not yet exist. They reflect a systems dynamic *approach. Systems dynamics can work on the basis of incomplete data. And data will always be incomplete. Many of the apparent paradoxes of a static system disappear in systems dynamics, and the hard-and-fast boundaries become fluid.*

Reviewing Part I as a whole, and viewing it in the context of the Pacem in Maribus project as a whole, the following points would seem to stand out for consideration:

1. The resources of the oceans are not static nor constant. They are what human ingenuity and technology make of them.

2. The possession of such technology is more important than the ownership of undeveloped resources.

3. Planning for the development of these resources must be systemic, *interlinking and harmonizing the multiple uses of the marine environment.*

4. Planning must be functionally, not territorially directed, in the context of natural, ecologic units, and according to ecologically determined geographical boundaries even though these do not corresepond to political boundaries.

5. Resource management and environment management must include the management of technology, i.e., a new science policy.

6. Resource management transcends the sphere of traditional international law and existing international organization and must be based on new forms of cooperation between government (national and international), science, and industry.

MILNER B. SCHAEFER

7 The Resources Base: Present and Future

BASIC GEOLOGICAL CONSIDERATIONS

In our consideration of the resources of the ocean floor, it is desirable to keep in mind the fundamental geological nature of the habitats of these resources, since this has a great deal to do with the nature of the resources and their distributions. The material presented here is largely abstracted from several recent papers.[1]

The two chief physiographic units of the earth are the continents and the ocean basins. The average level of the continents is some four kilometers above that of the ocean basins, because the continents consist mostly of lighter rocks. Put simply, the lighter continents are "floating" on the heavier material making up the ocean basins. The material of the continental crust, the so-called silicic layer, includes bedded sedimentary rocks within which oil, gas, coal, and other deposits are to be found, as well as the less dense crystalline rocks, such as granites, within which associated metallic minerals may occur as vein deposits and disseminations. The ocean basins, on the contrary, are made up of basic magmatic rocks, the so-called simatic layer, in which we expect to find minerals that are genetically associated with oceanic types of basic and ultrabasic magmatic rock, for example, chromite, nickel and platinum. Most of the ocean basin, however, is covered with

MILNER B. SCHAEFER, who died in August 1971, was the director of the Institute of Marine Resources at the University of California in San Diego.

a thick layer of sediment so that access to the underlying basement rock, and its contained resources, is very difficult.

The continental margin, the submerged edge of the continental block, is a submarine apron that includes the shelf, or shallow platform, the continental slope beginning at the outer edge of the shelf and going toward the depths at a sharper angle, varying from as little as three degrees to over forty-five degrees (twenty-five degrees being common), and the continental rise that is a broad, uniform, smooth wedge of elastic sediments that, wherever deep sea trenches are absent, slopes gently oceanward from the base of the continental slope in depths of two thousand to five thousand meters of water. The rise is composed, as noted, of a thick layer of terrigenous sediments, up to ten kilometers thick, overlying the basement rocks below.

In some places, such as the continental margin off Southern California, a continental shelf as a shallow platform does not exist, the seabed being broken up into a series of ridges and deep troughs, similar to adjacent continental land forms. This is known as a "continental borderland." It is, however, made up of the same kind of materials as the continental shelf and slope.

The boundary between the rocks of the continents and the ocean basins appears to underly the continental slope, or the inner part of the rise, but the exact nature and location of this boundary is not well-known. However, the minerals, sediment-types and structures of the continents and ocean basins are sharply separated at or near the base of the continental slope. Various depths have been suggested as a reasonable and practicable boundary, lacking more precise information about the details of rock and rock structure. Emery [2] has suggested one thousand meters for this purpose. Worzel [3] on the basis of careful study of existing geophysical information along various sections crossing the continental margin has stated that the edge of the continent is located approximately beneath the two thousand meter isobath, while 2,500 meters has been suggested by Dr. William Pecora of the U.S. Geological Survey. As may be seen from the compilation of the hypsometry of ocean basin provinces by Menard and Smith [4] (see also McKelvey and Wang [5] and McKelvey, et al.[6]), any of these bathymetric contours leaves a large portion of the seabed beyond the boundary; the percentage of the ocean floor included within the one thousand meter, two thousand meter, and 2,500 meter contours are, approximately, 11.9, 16.3, and 20.5, respectively, while that within the two hundred meter contour is 7.5. Thus, at least eighty percent of the seafloor is associated with the ocean basins rather than with the continental blocks.

About half the deep seafloor is covered with abyssal plains and hills, that lie at depths of about three thousand to 5,500 meters, consisting of relatively flat to rolling and hilly plains, studded with seamounts largely of volcanic origin. However, in some areas, the abyssal plains and hills have a rugged surface as the result of extensive fracture zones and faults. The cover of unconsolidated sediment is generally less than one kilometer in thickness, but thicker accumulations may be found locally in some areas. Underlying rocks consist predominantly of basalt.

About forty percent of the deep seafloor consists of oceanic rise and ridge, that is oceanic mountain ridges and slopes rising one thousand to three thousand meters above the adjoining abyssal plains and reaching the ocean surface in some places as volcanic islands. Along the mid-ocean ridges, that run generally through the major oceans, there is commonly a rift valley at the ridge crest, bordered by high ridges offset along numerous transverse fractures or faults which also cut the adjacent slopes. These mid-ocean ridges are believed to be areas where material from the earth's mantle is moving upward and spreading outward to create continuously new deep seafloor. Much of the ridges and oceanic rises is underlain by bare rocks, largely basalt, but a thin veneer of sediment such as red clay or biogenetic sediments (oozers) is present in some areas, and along the flanks of the oceanic rise thin sequences of older sedimentary rocks sometimes occur.

Common in the ocean basins are islands, banks, ridges, guyots, and seamounts composed of basalt of volcanic origin. Some of these are capped by a smooth platform with sedimentary deposits on them; others have been capped by coral growths.

Adjacent to the convex sides of island arcs, or along tectonically active coastal mountain ranges, commonly occur deep ocean trenches. Most of these lie around the margin of the Pacific Ocean, but a few, such as the Puerto Rico and Sunda Trenches, are on the edges of the Atlantic and Indian Oceans. These trenches include the deepest parts of the seafloor, and their floors are generally at depths greater than six thousand meters.

Volcanic islands, banks, etc. account for only some three percent of the seafloor, and the trenches and associated ridges for less than two percent.

MINERALS AND CHEMICALS

In considering the mineral and chemical resources of the sea (exclusive of oil and gas, that will be discussed separately below) it is convenient to consider them in the following categories: [7] (1) Mineral deposits within bedrocks, or vein deposits. (2) Surficial deposits. These are of two kinds, the placer deposits including such things as tin, gold, diamonds, iron sands, monazite and other such resistant or heavy minerals, that are deposited in the continental shelf areas; and chemical precipates, including especially phosphorite that is deposited by precipitation from seawater on certain areas of the seabed, and the ferromanganese nodules, precipitated from seawater mostly on the abyssal seafloor. (3) Minerals dissolved in seawater. (4) Metalliferous brines and muds.

Although the ocean, and even that portion beyond national jurisdiction, is of potential importance as a source of minerals and chemicals, it is desirable at the outset to dispel the widespread idea that the ocean will, within the foreseeable future, supply some large portion of the world's mineral needs. Harold James [8] has concluded:

> The mineral resource potential of the deep ocean is small per unit area compared with that of the continents. The chief reasons for this are (1) absence of a thick sialic crust, within which ore-producing magmas of granitic composition are generated on the continents; (2) no rocks older than Cretaceous are known to be exposed in the deep ocean, whereas the structural dynamics and erosional processes of the continents have resulted in extensive exposure of ore bearing Precambrian and Paleozoic strata; (3) important sedimentary and residual deposits such as evaporites, iron-formation, bedded phosphate, placers, coal, and laterite, either cannot form in the deep ocean or are highly unlikely.

He has also noted that:

> The mineral potential of the mid-ocean ridge, the greatest mountain system on earth, is very low. Emergent parts, such as Iceland and Hawaii, consist almost entirely of basalt that contains no economic mineral concentrations.

Bedrock or vein deposits

So far as this kind of deposit is concerned, there is to be recognized an important division between the two fundamentally different geological environments.[9] First, there are the geological formations of the continents that generally don't extend beyond the base of the continental slope and are contained within the continental crust. These include the bedded sedimentary rocks in which occur petroleum and gas (that will be discussed below) and also sulfur, coal, and bedded salt and potash deposits. In the light crystalline rocks, such as granites, we find associated metallic minerals, such as gold, tin and copper, as vein deposits and disseminations.

These deposits on the continental margins are, essentially, the same kinds as those on the emergent portion of the continent, and whether or not they can be economically exploited depends a great deal on the relative costs of terrestrial and marine mining and transportation to markets. At the present time, some coal deposits beneath the sea that are extensions of those on the land are mined by tunneling from shore, for example off Chile. Some sulfur is being produced, by the Frasch process, where it occurs in salt domes in relatively shallow water, as in the Gulf of Mexico. Cloud [10] has called to our attention that, although the substructure of the continental shelf and slope has comparable mineral deposits to the rest of the continent, there are certain problems peculiar to the region. Bedrock is more likely to be blanketed by post-submergence sediments on the continental shelf than on the eroding elevated lands, in consequence of which, in addition to the difficulties with the overlying water, we can't expect to find as high a proportion of the mineral deposits at the surface. In consequence, it seems very unlikely that within the foreseeable future such bedrock deposits beneath the sea, and especially beyond national jurisdiction, will be of any great importance.

The second important division of the bedrock deposits are those expected in the abyssal depths of the ocean, beyond the continental margin, in which should occur minerals that are genetically associated with oceanic types of rock, such as chromite, nickel, copper, and platinum. It is to be noted that iron and magnesium are probably the most abundant metal elements in the oceanic rocks, but these rocks could not compete with iron sources from dry land, or magnesium from seawater (see below). Further, beyond the continental margin difficulties increase, because much of the bottom is blanketed by thick sediments, and we can seek outcrops only on the seamounts and along the mid-ocean rises. But both James [11] and Cloud [12] are of the opinion that these prospects are not very promising, because the modern theory of seafloor spreading implies that, beneath a thin veneer of sediments, the ocean basins are generally floored with relatively young and sparsely mineralized basaltic rocks, and the rocks along the mid-ocean ridges are the youngest of all.

Surficial deposits—placers

Interestingly enough, among the more important nonliving products from beneath the sea are sand, gravel, oyster shell, and limestone, used for construction purposes. These all come from the very shallow margin of the nearshore zone, although some of them, particularly sand and gravel, also occur rather further offshore. However, the nearshore resources are undoubtedly more than adequate for the foreseeable future, and there is essentially no prospect of these being mined competitively at depths greater than two hundred meters.

Beaches and submerged placer deposits are already of importance for some kinds of heavy minerals, diamonds, etc. Placer deposits now offshore were originally formed by gravitational segregation during transport in and beneath former beach and stream deposits, when the sea stood at lower levels or the land at higher ones.[13] The general outer limit of depth where one expects such deposit is about 130 meters, the position of the beach during the last ice age, although it may be somewhat greater or less where the land itself has moved vertically. Currently diamonds, gold, and tin are being recovered from nearshore submarine placers, and the production of these is expected to increase, along with zircon, feldspar, rutile, ilminite, and other materials that are presently being produced from marine beaches and occur also in somewhat deeper waters.[14] Certain types of iron bearing sands are being exploited in a few locations on the continental shelf, and others will certainly be exploited in the not distant future. However, all of these things occur almost exclusively in relatively shallow water on the continental shelf, within the limits of national jurisdiction, so need not concern us in our present consideration of a regime for the resources of the sea beyond those limits.

Minerals dissolved in seawater

As has been frequently pointed out, seawater contains dissolved in it a great many elements, and although only a few of them are highly concentrated, the tremendous volume of the ocean contains large total amounts of even the trace elements. It is noted, for example, that there are ten billion tons of gold in seawater (although it represents only six parts in 10^{12} parts of water). Solar evaporation of seawater, in suitable locations around the world, to obtain sodium chloride, potassium chloride, magnesium oxychloride, etc.) is an ancient industry. The winning of magnesium metal from seawater, by precipitation with calcium carbonate and electrolytic processing, is the most important source of that metal in the United States. Chlorine and bromine are also economically extracted from seawater on a large scale. We must also remember that the ocean is, in appropriate circumstances, an excellent source of fresh water; desalination of seawater is already economical for some domestic and industrial uses in very arid coastal zones.

It has recently been shown that, by coupling a large-scale desalination plant with a chemical processing works, it may be possible economically to obtain, in addition to those now extracted, a few more elements from seawater, including boron, aluminum, lithium and fluorine.[15]

It is also to be noted that some of the trace elements, such as uranium, may prove to be profitably extractable by special processes.[16]

We need, however, for our present purposes to spend little time considering an international regime in relation to the extraction of materials dissolved in seawater. This is because the relative proportions of the major ions in seawater are very nearly constant in all parts of the world ocean, and the total salinity in the open sea varies by only a few parts per thousand. Thus, off every open seacoast there are large quantities of "ore" immediately available and constituting essentially a free good. The locations of plants for extracting materials dissolved in seawater depends, in consequence, on other factors, such as suitable land and climate for solar evaporation, quantities of calcium carbonate and of cheap power for extraction of magnesium metal, quantities of cheap power for extraction of bromine and chlorine, availability of nearby markets, transportation costs, etc. It is beyond any reasonable possibility that these factors of production will be less expensive anywhere in the ocean beyond the limits of national jurisdiction than they are on the seacoast, under national jurisdictions.

Metalliferous brines and muds

Recent discoveries of metalliferous brines, and metalliferous muds, in the deeps of the Red Sea, and elsewhere along rift zones or in areas of vulcanism, give reason to believe that such deposits may exist in numerous locations in the ocean floor.[17] The "hot holes" in the Red Sea contain brines with zinc, copper, and other minerals in concentrations ranging from one thousand to fifty thousand times that of normal seawater, and the muds in these areas in the bottom of the Red Sea are rich in copper and zinc. Iron and manganese precipitates have been reported from a submarine volcano in Indonesia, and sediments containing five percent manganese and one tenth of a percent copper have been encountered in the rift zone of the East Pacific Rise. The occurrence of metalliferous hydrothermal brines in California near the Salton Sea, where there is a terrestrial continuation of the rift zone along the East Pacific Rise, gives further reason to believe that similar phenomena may be encountered in such places as the Gulf of California, and elsewhere.

It has been estimated [18] that the upper ten meters of mud in one of the "hot holes" in the Indian Ocean contains 2.9 million tons of zinc, 1.1 million tons of copper, and smaller amounts of other metals. It is my understanding that, in this case, jurisdiction is claimed by two of the adjacent coastal states.

We obviously have yet too little information about these kinds of deposits in the deep seafloor, beyond the limits of national jurisdiction, to do much useful planning towards their exploitation and management; it seems likely, however, that some of the submarine metal-bearing muds may be economically minable within one to a few decades.

ENERGY RESOURCES (PETROLEUM AND GAS)

Thermal energy from fossil fuels has been the mainspring of development of modern industrial civilization. World production of thermal energy from coal and lignite plus crude oil has risen exponentially from about 5×10^{12} kwh/yr at the turn of the century to about 37×10^{12} kwh/yr currently [19] and over half of this is now supplied by petroleum. There are numerous forecasts of world energy requirements, and of world petroleum demand. All of them indicate that, for the remainder of this century, the supply of petroleum and natural gas will be extremely important, and perhaps critical. As mentioned below, there are possibilities of substituting to a degree the vast reserves of coal, and of obtaining oil and gas from tar sands and oil shales, while nuclear energy is developing apace. However, the situation has been well summarized by Weeks,[20] who has estimated that:

> Over the next twenty years the world will consume something like 500 billion bbl of petroleum and 750 Tcf of natural gas. These amounts are about 110% and 75%, respectively, of the current world proved reserves of these sources of energy. Since oil and natural gas consumption twenty years hence will be about four times that of today, the petroleum industry will be called upon to find several times 500 billion bbl of oil to replace that consumed and still maintain a safe inventory.

The development of petroleum fields beneath the sea has been remarkably rapid in recent years. The industry had barely gotten its feet wet in 1946, while offshore production now accounts for about seventeen percent of total world production of petroleum. Between nine thousand and ten thousand wells have been drilled offshore, and production is coming from as far offshore as seventy miles and in water over one hundred meters deep. Twenty-eight countries are already producing or about to produce subsea oil and gas. It is anticipated by Weeks [21] that by 1977 the offshore fields will provide thirty-three percent of an anticipated production of 25.5×10^9 bbl.

To put these requirements in perspective, it has been estimated [22] that present proved and probable reserves of world crude oil are in the neighborhood of 600×10^9 bbl and that ultimate recovery of petroleum is about $2{,}000 \times 10^9$ bbl. One of those responsible for the latter estimates is Weeks, in an earlier paper, but in his most recent publication [23] he comes up with an estimate of ultimate recovery of $1{,}500 \times 10^9$ bbl of petroleum on land and 700×10^9 bbl of petroleum offshore to a depth of one thousand feet (256 meters); he also estimates that the natural-gas resource is the equivalent of 800×10^9 bbl of oil on land and 350×10^9 bbl of oil in the offshore region, again to one thousand feet water depth.

Weeks notes that the area out to one thousand feet is 10.8 million square miles, and of this thirty-seven percent or 6.2 million square miles, is sedimentary basin where one can consider looking for petroleum, and that this is one third as large as the 18.5 million square miles of the world's land basins.

It is to be remembered that an alternative to petroleum and other fossil fuels is nuclear energy, and this is beginning rapidly to be developed commercially. However, present successful commercial applications of nuclear energy depend on fission of enriched uranium fuel, which is both limited in ultimate supply and costly. Great success in providing low-cost nuclear energy appears to lie with the development of the fission breeder reactor, and, ultimately, the fusion reactor. While power from nuclear fission, together with petroleum from oil shales or tar sands, and conversion of coal to liquid hydrocarbons, offer sufficient economic competition to petroleum and natural gas to limit acceptable increases in production costs, it is almost certain that natural deposits of petroleum and gas, including those in at least the shallower portions of the seabed, will continue to satisfy a major share of the world's energy requirements during at least the remainder of this century, and doubtless beyond. However, Weeks [24] states that the combined totals of potential synthetic oil and gas from shale and coal sources is many times the quantities indicated above as the estimated potential for direct production.

Marine habitats of petroleum and natural gas

As the review of the U.N. Secretary-General [25] has pointed out, the origin of offshore petroleum, and the factors controlling its distribution in sedimentary basins in the shelf and slope, are no different from those on land. From the viewpoint of the genesis and accumulation of hydrocarbons, the most important aspects are petroleum source beds with abundant organic matter, reservoir rocks, structural stratigraphic traps, and geological history, particularly sedimentation and structural development. The petroleum hydrocarbons arise from organic matter from organisms growing in the sea that, after death, are transformed by bacteria and eventually buried beneath the detrital and biogenic sediments where hydrocarbons are generated. Under suitable geological conditions, these petroleum hydrocarbons migrate into structural or stratigraphic traps, where they may be exploited commercially by drilling of wells.

Large commercial deposits of petroleum are known to exist on the continental shelf, both from geophysical evidence and from actual development of oil fields there.[26] Some of the offshore regions now in production are seaward extensions of onshore petroleum deposits, but another group of petroleum accumulations in the shelf are in structural stratigraphic traps that are separate from, but have a distribution pattern related to, similar structures previously discovered on land. The nature of the habitats of petroleum on the shelf and evidence concerning its probable occurrence on the slope and further offshore have been discussed by a number of authors.[27] The situation is well summarized by McKelvey and Wang [28] as follows:

> Petroleum resources are largely confined to the continental shelves, continental slopes, continental rises and the small ocean basins. Because these areas in general contain a greater thickness of marine Tertiary sediments, from which most of the world's petroleum production comes, than do the lands, taken as a whole the offshore areas are more favorable for petroleum than the exposed parts of the continents. Environments favorable for petroleum are highly localized; and . . . only a small part of the broadly favorable areas actually contain producible petroleum accumulations. . . . Among the geological provinces considered broadly favorable, the incidence of petroleum accumulations in the shelves, slopes, and the small ocean basins may be greater than in the continental rises bordering the large ocean basins. Although the rises contain greater thicknesses of sediments, in many places they may not contain suitable reservoir rocks.

The National Petroleum Council [29] agrees that the semi-enclosed seas, such as the Gulf of Mexico, Black Sea, Caribbean, and South China Sea, often have thick sedimentary sections in deep water beneath their abyssal floors, so that they may be habitats of petroleum. However, the structural stratigraphic nature of their sedimentary fill is just now being learned. The Gulf of Mexico has certain salt domes (the Sigsbee Knolls) that have shown traces of hydrocarbons in cores recently drilled by the JOIDES project.

The National Petroleum Council [30] also notes the very thick sediments on the lower slope and continental rise, and the existence of some locations with favorable structural features that would argue for a favorable petroleum potential.

McKelvey and Wang point out that the area beyond the region covered by the estimate of Weeks, that is from a depth of about three hundred meters to the toe of the continent, constitutes a larger area than that included within his estimates, and, furthermore, a larger proportion of the area is underlain by a thick accumulation of sediments. However, until more is known about the composition and structure of these sediments it is not possible to judge their potential, although it is likely that they are quite good.

So far as the floor of the great ocean basins is concerned, that is the sediments of the abyssal seafloor, both the report of the U.N.

Secretary-General [31] and the report of the National Petroleum Council [32] indicate that any sizable deposits of hydrocarbons are unlikely. The National Petroleum Council observes that knowledge is quite limited, and that hydrocarbons probably occur in at least trace quantities in the sediments of almost all areas of the ocean, but that current information suggests that commercial accumulations will be far fewer per unit area beyond the continental margins. The U.N. Secretary-General's Report is even less optimistic, stating that:

> All evidence suggests that the abyssal open oceans are far less favorable than the continental margins and small oceanic basins and there is little chance that petroleum occurs over large areas of the abyssal plain. However . . . no definite seaward limit for the existence of petroleum deposits can be inferred at this time, and it is not impossible that small portions of the abyssal floor and oceanic trenches may have some potential.

Prospective rates of development

Current offshore petroleum production comes from water depths with a maximum of only a little over one hundred meters, and from areas within 120 kilometers of the coast.[33] However, this is not so much limited by exploration technology, nor by capability of drilling in deep water, as it is by other production costs.[34] Indeed, there are actually many advantages in offshore exploration, because the application of geophysical seismic reflection techniques is much simpler there, and recent development of non-explosive sound sources has not only reduced the costs of seismic exploration but has greatly diminished the danger of damage to fish and other living resources.

So far as drilling technology is concerned, the industry is already drilling exploratory wells in water as deep as thirteen hundred feet (396 meters).[35] Indeed, the *Glomar Challenger,* which is doing scientific exploratory drilling through the sedimentary column in the abyssal depths of the ocean for the JOIDES group, is capable of doing so at almost any depth of water. However, ability of *Glomar Challenger* to drill at great depths through very hard structures is limited by inability, so far, to accomplish re-entry of the drill hole thus making it possible to replace worn bits. It is expected, however, that this capability will be developed within the next few years. According to the National Petroleum Council,[36] technology will allow drilling and commercial exploitation in water depths up to fifteen hundred feet within less than five years, and within ten years technical capability to drill and produce in water depths of four thousand to six thousand feet (1219 to 1829 meters) will probably be attained.

However, just how fast the development of deposits in deeper water, and in water farther offshore, will occur depends very much on economic factors. Costs of drilling, and of all other activities connected with petroleum production, increase very rapidly with increasing water depths and distances from shore.[37] Indeed, it was recently reported that gas discoveries in the Norwegian sector of the North Sea, beyond the deep trench off Norway, are not economical because of the cost of getting the product to shore in Norway; it may be marketed in the United Kingdom.

Handling the petroleum and gas in great depths and far offshore, and getting it to shore, will be technically troublesome and costly. Consequently, just how rapidly the production of petroleum and natural gas from the seabed will progress out and down the continental slope will depend upon how rapidly these various problems are solved, and at what cost.

As already noted, the competition from other sources of petroleum and natural gas, such as oil shale, sets a limit on the costs that can be borne competitively. For example, oil in the United States is worth something like three dollars a barrel, although it is worth only about half that in many foreign areas such as the Persian Gulf. It is expected that oil shale from the vast deposits in Colorado, Utah and Wyoming will be capable of producing petroleum for about $2.12 per barrel by 1976 and $1.58 per barrel by 1980.[38] Also, as already indicated, this source of energy faces potential strong competition from nuclear energy, which tends again to put a ceiling on the costs that can be borne by offshore petroleum development.

Implications for an international regime

As we have seen above, it is highly likely that the major portion, at least, of petroleum and natural gas of the seabed occurs on the Continental Margin (on the Continental Shelf, the Slope, and perhaps the Continental Rise). Prospects are poor on the abyssal seafloor. Consequently, the question of just where national jurisdiction ends seems to be one of the most critical questions to be answered before one can forecast how soon a regime is required for the area beyond national jurisdiction. Should it be limited, as some have suggested, to a depth of two hundred meters, it is to be observed that some leases have already been let by more than one nation, and production will shortly be taking place, beyond that depth. On the other hand, should the limit of national jurisdiction include, as some authorities have asserted that it does already, at least potentially, the entire continental terrace,[39] there would seem little need for an international regime for yet some years.

However, whatever regime governs the exploitation of subsea petroleum and natural gas, it will need to be such as to guarantee to the entrepreneur security of tenure to a reasonably large area for a reasonable length of time because of the highly localized and highly concentrated nature of the deposits to be exploited and because of the large investment in the equipment for extracting it from the earth and storing it locally as well as for transporting it to shore by vessels or pipelines. T.F. Gaskell [40] has discussed a number of these factors at some length. He concludes that, even on the continental shelves, it is necessary, at present-day values of petroleum, to find very large reservoirs to make any discovery an economic one, and he further points out that any hopes of a world organization becoming rich by having authority over the oil and gas beyond the continental shelf are unrealistic.

It will also be necessary for the regime to encompass some means of insuring against undue interference with other uses of the seabed and the overlying waters, and especially pollution that can damage the living resources. Amelioration of dangers of pollution also needs to be encompassed in the case of other types of seabed mining, as will be discussed further below.

LIVING RESOURCES

Despite rapid increases in recent years in extraction of petroleum and other non-living resources of the sea, the extraction of living resources by fisheries throughout the World Ocean is still the major leading activity. According to the most recent tabulations of FAO,[41] the harvest of the marine sea fisheries in 1968 was 57.3 million metric tons. The value of this harvest, at the fishermen's level, is roughly double the value of petroleum and minerals extracted from beneath the sea.

Although sea fishing is an ancient industry, the harvest has increased rapidly since shortly after World War II, having increased from 21.9 million to 57.3 million metric tons from 1952 through 1968, representing an annual average rate of growth of 6.2 percent which, of course, is a good deal faster than the rate of increase of human population. While this is a very satisfactory rate of growth, some authorities [42] believe that it will decrease in the near future. I will discuss the potential harvest and probable rates of growth further below, but it is well to set forth at the outset that the assertion, "Food production from the oceans may be tripled or quadrupled during the next few years" [43] is improbable in the extreme.

The fishery harvest of the sea is, and will certainly continue to be, however, a very important source of food for mankind. It is an important source of high quality animal protein, although its present or future contribution to total calorie requirements is almost trivial. As I have pointed out elsewhere, and has been discussed by many other authors, careful examination reveals that the sea is a poor place to look to for total food requirements of any large sector of the human population, because of the fact that, although the total fixation of organic carbon is roughly the same in the sea as on the land, the plants of the sea are almost entirely microscopic organisms with fast growth rates, short longevity, and small standing crops and so are not amenable to economic harvesting or to culture. Our food energy, calories, comes in large part from plants, and the sea is a poor place to harvest them. The food harvest of the sea consists, and for the foreseeable future will continue to, of animals, one or more steps above the plants in the food chain, that supply the high quality proteins containing the essential amino acids required for people's health and well-being. Already some fifteen percent, or more, of the world supply of animal protein comes from the fisheries. The harvest is increasing, as we have seen, and the potential is still large.

The living resources of the sea are so totally different in nature from its nonliving resources, especially those of the seabed, that a regime for their appropriate management and conservation must necessarily also be in many respects quite different.

In the first place, the petroleum and minerals of the seabed are "stock resources," that is, there is a certain stock of them, which may be used rapidly or may be used slowly, but the total amount available is fixed. The management, or conservation, problem for such stock resources consists in determining the distribution of use over time of the total fixed quantity that is most beneficial to man. The living resources, on the contrary, belong to a different class of resources, the "flow resources," the supply of which is constantly renewed. Among the flow resources, living resources have the unique and important property that the rate of renewal depends on the amount of the resource left to perpetuate itself, which in turn, depends on the rate of harvesting.[44] In consequence, for a population of harvestable organisms in the sea, as exploitation grows, the sustainable annual harvest increases only *up to a certain point,* and beyond that it diminishes. There exists then, in general, for each living resource a maximum sustainable yield, and this yield may be taken indefinitely, year after year, in perpetuity. Maintaining each population of living resources in the sea in that condition where it is capable of yielding the maximum sustainable harvest has been adopted by the international community as the appropriate criterion of conservation, this being defined in the Convention on Fishing and Conservation of the Living Resources of the High Seas [45] as:

> . . . the aggregate of the measures rendering possible the optimum sustainable yield from those resources so as to secure a maximum supply of food and other marine products.

There has been considerable urging from a number of economists, however, that a more appropriate criterion would be some lesser level of harvest, where the net economic yield is maximized, and some of them have advocated both national and international regimes designed to attain that objective.[46]

The living resources of the ocean also differ from the resources of the seabed in that they are not fixed in location, but are highly migratory. Indeed, even the so-called sedentary organisms, that are regarded under the Convention of the Continental Shelf as being resources of the shelf, often have larval stages that swim freely in the superjacent waters.

Many species of fish travel great distances. For example, the bluefin tuna of the North Atlantic migrate completely across that ocean. Likewise, the albacore tuna of the North Pacific and the bluefin tuna of the North Pacific have been shown to migrate from California to Japan and return. The case of the European eel, which lives much of its life in rivers in Europe, but spawns in the deep ocean, in the Sargasso Sea, is well-known. In addition to these spectacular, very long migrations, there are important lesser inshore and offshore movements of many species and, indeed, these are, in many cases, a necessary part of the life history of the species. For example, the penaeid shrimp that support major tropical fisheries spawn in the open ocean but the larvae migrate into estuaries and lagoons where the juveniles grow for several months before returning again to the open sea. Likewise, menhaden, herring, and some other clupeoid fish have life histories that involve inshore or estuarine waters during parts of their life cycles. The anadromous fishes, such as salmon, shad, and striped bass not only come into inshore marine waters to reproduce, but actually ascend rivers, some salmon going hundreds of miles upriver.

In consequence, it is obvious that the living resources of the sea do not recognize such man-made lines as the boundary of internal waters, the territorial sea, or the contiguous fishing zone. At some stages, or seasons, they may be in the high seas, well beyond the region of national jurisdiction, while at other times they may be well within national jurisdiction, even in internal waters.

A third important characteristic of the living resources of the sea is that the various different species are not completely independent entities. Each living resource that man harvests is a member of a complex ecological community, and its harvesting, in some cases, significantly affects not only the harvested population but also populations of its competitors, predators, and prey. Any adequate regime for the management of living resources of the sea, regardless of in what jurisdiction they may be found, must take these factors into account. In other words, to be successful, planning and management of the harvesting of the living resources of the sea must be according to units that reasonably correspond to natural ecological units, even though these may sometimes be politically inconvenient.

Magnitude of the potential harvest of the sea

Attempts to estimate the potential fishery harvest of the world ocean have been numerous in recent years. Schaefer and Alverson [47] have summarized many of the published estimates through 1966. In addition, Ricker [48] has recently made a careful estimate, and the global summary result of the FAO's study, part of the Indicative World Plan for Food and Agriculture, has been announced.[49]

Some of the early estimates, of the order of fifty-five to seventy million metric tons were obviously too low, since, as noted above, we have already reached fifty-seven million metric tons, and it's quite easy to see where another thirteen million tons may come from. Remaining estimates, however, vary all the way from figures in the neighborhood of 150 to two hundred million metric tons to values as large as one thousand or two thousand million metric tons. However, careful examination of the various papers in which these estimates have been published reveal that they are not as diverse as might appear at first blush. The lower values, of the order of two hundred million metric tons per year, are estimates of the sustainable harvest of organisms of the kinds that are presently supporting the world's fisheries, the increased harvesting of which will not require any radically new technology. The higher estimates, on the other hand, are based on large-scale harvesting of organisms smaller, and lower in the food chain, than most of those presently being harvested. Whether and when this will become economical is a moot question.

Two approaches may be taken to estimating the potential harvest of the sea. The first is to start with the total fixation of organic carbon by photosynthesis and, assuming ecological efficiency factors or transfer factors from prey to predator up the food chain, to arrive at estimated harvestable outputs at various trophic levels. The second way is to consider systematically what we know about various unused harvestable populations of fish, squid, shrimp, and other marine organisms as a basis of estimating how much increased harvest might be attained. Employing the former, tropho-dynamic approach, Schaefer [50] calculated, for a primary phytoplankton productivity of the world ocean of 1.9×10^{10} tons of carbon per year, the potential yields at various trophic levels, for a series of assumed ecological efficiency factors. On that basis, and considering the approximate trophic levels at which the harvest is presently being taken, he concluded that two hundred million metric tons per year is probably a conservative estimate for the potential harvest of the sea fisheries. He also emphasized, as have other authors, that as one goes into lower trophic levels, the harvest can be greatly increased.

The recent estimate of Ricker,[51] referred to above, is based on a similar sort of calculation, but he assumed a primary productivity of only 1.3×10^{10} tons of organic carbon fixation per year, based on a 1965 paper of Koblentz-Mishke.[52] He arrived at a potential yield estimate of 150 to 160 million metric tons per year. Koblentz-Mishke, et al. have made an improved and revised estimate in a paper still in press,[53] obtaining an estimated rate of primary productivity of 2.3×10^{10} tons of carbon per year. It is interesting to note that, if one puts this new value for primary production into Ricker's calculations, he obtains in place of the 150 million metric tons of annual potential fishery harvest, 265 million metric tons.

Schaefer [54] also reviewed in a rough fashion what is known about unutilized fisheries resources in various parts of the world ocean, and concluded also from that approach that his estimate of two hundred million metric tons was reasonable. More recently, the FAO has conducted a painstaking, area-by-area survey of the world ocean as to what is known about unused fishery resources and, on that basis, according to a report of a recent FAO conference on fisheries investments [55] has arrived at a potential world harvest of about 120 million metric tons per year.

The discrepancies among these several estimates reflect the fact that our data, using either approach, are very poor. It is quite evident, however, that the present sea fishery harvest can at least be doubled and probably quadrupled without any far-out new technology.

The foregoing estimates have assumed no large-scale development of fisheries for organisms at lower trophic levels presently unharvested. Should it prove economical to exploit these lower trophic levels, harvesting such things as the Antarctic krill *(Euphausia superba)*, or other euphausids, and such things as myctophid fishes and deepsea smelt, the total catch may be further greatly increased. For example, based upon the amount that is estimated to have been formerly eaten by whales in the Antarctic, it appears that something like fifty to one hundred million metric tons per year of *Euphausia superba* alone could be harvested.

In considering the global harvest of the World Ocean, it is important to emphasize the fact that the fish populations, and the obtainable harvest from them, are not uniformly distributed over the World Ocean, which has its deserts and its green pastures. The highly productive portions of the ocean are those where plant nutrients are renewed from deep water to the sunlit zone of the sea to support the growth of the phytoplankton upon which all else depends by stirring up from shallow bottoms, by mixing along current boundaries, by winter overturn at high latitudes, and by coastal upwelling.[56] Although there are highly productive, far-offshore areas of the World Ocean, such as the upwelling zones along the equator or the areas of winter overturn at high latitudes, in general the near-shore portion of the ocean is the most productive, both because of shallow bottoms from which the nutrients may more easily be returned to the surficial waters and because of the existence of strong coastal upwellings. In consequence, although there are highly productive fisheries in mid-ocean in some places, a major portion of the fisheries harvest is taken near the margins of the continents, and major unutilized resources are known to exist there, a notable example being the western side of the Arabian Sea. The highly productive zones near the continental margins usually extend much further offshore than the territorial sea, or the contiguous fishing zone (usually twelve miles wide), recently claimed by many nations. This matter is very important, because this distribution of the fisheries, together with the economic dependence of some coastal communities on nearby fisheries, supports to some degree claims to certain preferential rights of the coastal state in the fisheries off its shores, even on the high seas, which were first formally recognized in the Convention on Fishing and Conservation of the Living Resources of the High Seas.[57]

Prospective rate of development

As I have noted above, the annual average rate of growth of the world's marine fisheries during the last seventeen years has been 6.2 percent. The growth rate from 1952 through 1957 was 4.6 percent, during the next five years was 7.9 percent (this was the period of development of the anchovetta fishery off Peru, that has attained its maximum sustainable harvest level of about ten million metric tons per year), and from 1962 through 1967 it was 5.5 percent (from 1962 through 1968 it was 6.2 percent). I see no reason why a rate of growth in the neighborhood of five percent to six percent a year should not continue, since there are certainly adequate resources to support this for yet a good many years. On the other hand, Ricker [58] writes that the present rapid rate of increase is unlikely to last much longer, and believes that if the world reaches one hundred million tons by 1990, that will represent very good progress (I note that this would be an average rate of growth of only 2.5 percent per year from now until 1990). Likewise, many persons attending the FAO Conference on Investment in Fisheries [59] were concerned with the lack of prospective capital for wide-scale fishery development. While I respect these opinions, I cannot agree with them, in the face of knowledge of large unexploited populations of fish along the western side of the Arabian Sea, in high latitudes of both coasts of South America, and other locations, and knowledge that adequate capital is available for immediate investment in fisheries for production of fishmeal and other commodities from the abundant resources, such as those on the western margin of the Arabian Sea, at such time as the political and investment climate becomes favorable.

While I believe that we can look forward to continuing growth, at a satisfactory rate, of the world sea fisheries, it is quite evident that the resources that support many of the major fisheries are at, or approaching, the level of maximum sustainable yield and that their conservation is of the utmost importance. As the sea fisheries grow, even more of the fish stocks will reach the condition where conservation rather than development becomes the major problem.

Some implications for a regime for international sea fisheries

From the foregoing it seems evident, at least to me, that in planning any regime for the rational utilization of the living resources of the World Ocean, the following factors must be taken into account:

1. The need to provide for management of the living resources by natural species-populations, in the context of natural ecological units, and according to their ecologically determined geographical boundaries, even though these do not correspond to political boundaries.
2. The need to encourage full development of underused resources and to provide for adequate conservation of those that are fully utilized, that is, as a minimum, to prevent exceeding the rate of fishing corresponding to the maximum sustainable yield.
3. The need to obtain adequate, objective scientific information concerning the biology, ecology, and population dynamics of the living resources to be managed in order to provide the basis of satisfying the two foregoing requirements.
4. The need to recognize the rights of coastal states and also the rights of distant-water fishing nations, operating in the high seas adjacent to the coast.
5. Where a resource is substantially fully utilized, the need to establish some acceptable basis of sharing among nations of the available harvest.

6. The need to provide for some adequate system of adjudication of disputes.

All of these factors were recognized at the time of the International Conference on the Law of the Sea in 1958, and, in my opinion, the resulting Convention on Fishing and Conservation of the Living Resources of the High Seas [60] provides a reasonably adequate framework for international management of the sea fisheries if nations would use it.

In the several conferences and other international deliberations preceding the 1958 International Conference on the Law of the Sea and at the Conference on the Law of the Sea itself, the idea of placing the sea fisheries under the jurisdiction of some single international authority (FAO or other) was specifically rejected. It was also recognized that the most successful international research and management programs had been those dealing with a species or a group of species in a specific region of the World Ocean. It continues to be evident that the regional international approach, involving either a particular resource or a group of resources or an entire oceanic region corresponding reasonably to an ecological unit, is the most fruitful way to satisfy item (3) above, which is the basis for satisfying needs (1), (2), (6), and, to some extent, (4).

The 1958 Convention also provides a reasonable and generally acceptable balance between the rights of coastal States and the rights of distant-water fishing nations and establishes an arbitration machinery for resolving disputes. Unfortunately, there is apparently little inclination among nations to utilize this machinery; the resolution of current disputes apparently depends mostly on direct political negotiations among the countries concerned. In at least some instances, such as the dispute between Chile, Ecuador and Peru and the distant-water fishing nations catching tuna off their shores, this direct political negotiation seems to have been a dismal failure. One would hope that any revision of the regime would establish some improved incentives for utilization of the 1958 Convention or other arbitration or juridical machinery for resolution of disputes.

The most important point not covered by the Convention of

1958, which has been the basis of much recent discussion, is the need to establish some acceptable basis of sharing among nations the sustainable harvest of a fully utilized resource. A great many formulae have been advocated, too numerous to review here, but none of them seem to be generally acceptable. In some few fisheries there has been *de facto* exclusion of all nations except a few participants, and an agreed sharing of the catch among the participants. The Antarctic whale fishery is a notable example, although in this case the whale stocks became so badly overfished that they were almost economically extinct before this was done. There is *de facto* agreed allocation of the catches of salmon in the northwestern Pacific between Russia and Japan. In the northeastern Pacific there is *de facto* exclusion of all salmon fishermen except those from the United States and Canada, an agreed sharing of the catch involving the allocation to Japan of a small portion of the catch of salmon originating in Bristol Bay that migrate westward of the agreed upon "abstention line." This, however, is apparently a somewhat unstable arrangement, because Korean fishermen who are not parties to the agreement are planning to enter the fishery. There have also been discussions toward some kind of agreed allocation of the catches of cod and haddock in the north Atlantic Ocean. This critical problem of acceptable allocation of the harvest seems yet to be far from a solution and is one of the more important considerations in the development of any revised regime for the international management of the living resources of the high seas.

HIROSHI KASAHARA

8 International Aspects of the Exploitation of the Living Resources of the Sea

BACKGROUND

The ocean contains a great variety of resources which could be utilized for the benefit of mankind. In terms of gross value to producers, however, the fish (as this word is used commonly, it also includes invertebrates and other non-fish animals) still remain the most important, valued at an estimate of some ten billion dollars a year. The only other thing which comes even close to the fish is oil (and natural gas). But the exploitation of seabed petroleum is in many ways an extension of oil production on land. Furthermore, the petroleum industry is so strong and self-sufficient that there is really nothing the international community can do for facilitating its development except dealing with legal matters. International legal problems arising from the exploitation of seabed petroleum has also been dealt with fairly satisfactorily through negotiations between the countries directly concerned. The exploitation of other mineral resources is, generally speaking, still at an early stage of development, and progress in this area has not been as rapid as many people thought a few years ago.[1] In short, the fish maintains its unique position among all the resources of the sea as far as its practical value to mankind is concerned.

Another aspect which characterizes the exploitation of marine organisms is that most of the nations in the world have long-

---

HIROSHI KASAHARA is the Associate Dean of the College of Fisheries, Bureau of Fisheries, University of Washington in Seattle.

established historical interests in the use of the sea for this purpose. The sea has always been an important source of food for man. It is also possible that fishing might be an industry even older than agriculture. The history of fishing activities on the high seas is not very new either. As early as the mid-1500's, the Portuguese were exploiting the cod schools of the Grand Banks of Newfoundland. The American whalers were all over the world in the eighteenth century. Trawl fishing in the offshore areas of the North Sea had been developed by the mid-1800's.

There have been two periods of massive fishery development in the modern history. The first one occurred during the late nineteenth and early twentieth century in connection with the Industrial Revolution. This was limited, however, more or less to European and North American nations and Japan. The second massive wave of development came after World War II, and we are still witnessing its world-wide effect. The overall rate of growth in the last ten or fifteen years has been quite impressive (roughly six percent per annum in total production). The period has also been characterized by the diversity of ways in which different nations have contributed to the world-wide expansion of fishing and related activities.

Peru, for example, suddenly became a great fishing nation (landing the largest volume of fish in the world) and the world's largest fishmeal producer by taking advantage of the presence of an enormous anchovy resource in her coastal waters. Japan, initially pressed by immediate needs to feed the nation and to increase foreign exchange earnings, not only intensified fishing in nearby waters but also developed mothership-type fisheries of various kinds, a tuna longline fishery to cover the tuna grounds of practically the entire World Ocean and trawl fisheries to operate in waters off West Africa and other distant areas. The Soviet Union, through state enterprises, has developed its distant-water fisheries in a very systematic way. A large number of vessels, including many gigantic ships with built-in factories, have been constructed not only in Russia but also in many other countries. Soviet trawlers have operated in virtually all major international fishing grounds. Many of the other European countries have expanded their fisheries by employing larger vessels and much-improved equipment.

In the field of fishery development, not only Peru but also a number of other developing countries have done rather well. In fact, the average rate of expansion has been greater in the developing nations than in the developed countries. There are several factors which have contributed to such a success. Absence of a rigid institutional framework, such as land tenure, and quick returns from investment, often in much needed hard currency, are among them. In Asia, for example, South Korea and Taiwan have not only expanded their coastal fisheries but also made great strides in participating in the world longline fishery. North Korea, too, seems to have developed modern fisheries in nearby waters. The Philippines have successfully adopted a modern purse-seine fishery and expanded coastal trawling. Thailand now has one of the largest trawl fisheries in Asia. Malaysia, too, is in the process of rapid expansion. Dramatic developments are beginning to take place in the rich and hitherto greatly under-utilized waters of the Indonesian Archipelago. India has developed a shrimp industry which is now the second largest in the world. In waters along the west coast of Africa, both fishing vessels from the African nations and those from European and Asian countries have participated in the development of enormous resources available there.

In addition to these and other remarkable examples of developments, the mechanization of small inshore craft along with the introduction of modern fishing gear in practically all of the developing countries has been a major factor contributing to the growth of fish production from the ocean.

Both scientists and industrialists in the fishery circle are now seriously concerned about future prospects of fish production from the sea. In view of the rapid rate of expansion of fishing and related activities all over the world, the question is now one of practical importance. How long can this trend of expansion continue? What new resources are likely to be found and exploited in the foreseeable future? What is the potential harvest of the sea?

Many of the forecasts made before 1962 were much too low and have been or are being exceeded by actual production. Scientists are taking another look at the questions on the basis of more recent information. Three methods are often used: extrapolation of the present trend, theoretical exercise based on productivity data, and estimation of the potentials of the individual resources, the existence of which are more or less known to us. Estimates of very wide range have been presented, depending on the methods used and assumptions adopted. It can be said, however, that the potential harvest of the sea is many times the present catch, and that, even without major breakthroughs in the technology of harvesting or marketing,[2] the world's fishery production may be further doubled in the next ten to twenty years.

But the importance and potential of food production from the sea must be assessed realistically. There is no evidence to support the often expressed hope that the problem of feeding the expanding world population could be solved by exploiting ocean resources. An overwhelmingly large proportion of the food requirement has been and will continue to be met by agricultural products.[3] The main contribution from the ocean will be to increase the quantities of animal protein products available for direct or indirect [4] human consumption.

The purpose of the above brief review is to point out that most of the nations have already real vested interests in the exploitation of living resources of the sea, that their interest and patterns of fishery development vary greatly, that food production from the sea has been increasing steadily and will continue to grow for some time to come, and that the developing countries have generally done rather well in this area. International aspects of the utilization of living marine resources should be considered against this background and should not be dealt with as new problems to which a simple and overall solution is to be found.

INTERNATIONAL REGIME

The regulation of fishing activities is a very complex matter even on a national basis. Most of the fisheries resources are made up of renewable wild stocks of animals which are highly mobile. They cannot be fenced in limited areas or marked for ownership. A great variety of usable fishes are found in the same body of water, and a number of species can be caught simultaneously by the same type of gear, while the same species may be fished by a number of different methods. Most of them undergo planktonic stages in which they are members of communities consisting of a much greater variety of organisms. Means to study the biological characteristics and population dynamics are limited and inaccurate compared with those used for land animals. The rate of renewal is rapid and little is gained by keeping them unexploited.

A legal framework for controlling the utilization of such resources is certain to be much more complicated and ineffective than that for terrestrial resources. In addition, there are problems arising from the heterogeneity of the industry. There still exist many fishermen who are catching fish with methods not so different from those used hundreds of years ago, while large fishing companies are dispatching their fleets all over the world.

Internationally, problems are even more complex. Fishermen from nations at greatly different levels of economic development may operate in the same fishing areas often using similar techniques and equipment. For example, the per capita gross national product (1965) among the nations fishing in West African waters ranged from sixty dollars to $3,240. The need for, and interest in, fishery development also differs greatly from nation to nation. Some nations use their fish resources mainly for producing products for export to earn foreign exchange. On the other hand, in a large number of nations, particularly those in East Asia and Southeast Asia, fisheries are very important for supplying the populations with animal protein food. Often roughly one-half of the total animal protein intake is from seafood. For many nations, it really does not matter whether or not they can develop fisheries, either because they have sufficient animal protein supply from other sources or because they can buy as much protein food as they need. The most systematic way in which a fishing industry has been developed is the pattern followed by the Soviet Union. Their determination to develop fisheries to increase animal protein supply has resulted in a tremendous investment in vessels, land facilities, and everything else that is needed by the industry. The impact of this is still being felt all over the world. Some of the

East European nations are following the example, though on a much smaller scale.

Because of the complexity of the problems involved and the historical background mentioned above, we still do not have a set of general principles for the regulation of high seas fisheries that are acceptable to most nations and are workable in specific situations. The four conventions which came out of the Geneva Conference in 1958 constitute only a very general framework as far as the regulation of fishing activities is concerned. The Convention on the Territorial Sea and the Contiguous Zone failed to include an agreement on the breadth of the territorial sea, or exclusive fishery zone. It has been a general trend in the past two decades for nations to expand their territorial seas or exclusive fishery zones. The bulk of the nations now claim twelve miles (either as territorial sea or a combination of territorial sea and exclusive fishery zone). Even if a nation does not officially recognize the twelve mile zone, it is becoming increasingly difficult for her nationals to fish freely within twelve miles of the coast of a state claiming it. A substantial number of countries claim zones much broader than twelve miles and up to two hundred miles.[5] I would assume that most nations still wish to keep some broad areas of the ocean as international waters outside of national jurisdiction, although more and more people are expressing pessimism about this. The number of nations claiming a zone broader than twelve miles and enforcing it is still comparatively small, and they are mainly in Central and South America. The trend of expansion of national jurisdiction, however, is definitely irreversible and would perhaps be accelerated each time the question is opened to discussion on a broad international basis.

The Convention on the High Seas is largely a codification of laws of navigation and the behavior of ships. The third convention, on the Continental Shelf, has considerable bearing on fishery matters. From the overall point of view of exploiting the continental shelf resources, the main problem is considered to be a lack of clear definition as to the outer boundary of the shelf. From the fishery point of view, it is a lack of clear definition as to what living resources should be subject to this convention. It appears, however, that this problem is being dealt with on a practical basis. The procedure being followed is for the member countries concerned with the utilization of a particular sea bed stock to determine, by consultations, whether they should be subject to the convention.

The fourth convention, on Fishing and on the Conservation of Living Resources of the High Seas, sets forth an obligation for member nations to enter into negotiations, if proposed, for an international agreement for conservation purposes. Although many of the major fishing nations are not parties to this convention, such a procedure for negotiating conservation agreements has long been followed in different parts of the world. It has generally been considered desirable to maintain the total physical yield from a stock at a high level. This general consideration has constituted a common ground for initiating negotiations for international agreements with a view to preventing decreases in sustainable yields from the resources concerned.

It is also assumed that most nations have diplomatic reasons for trying to avoid or minimize international disputes over fishery matters.

These general agreements or understandings are quite insufficient to form a framework in which specific problems of high seas fisheries can be dealt with. What has played a major role in the handling of international fishery problems is the conclusion of conventions for specific resources and areas through negotiations between the nations concerned and the operation of international bodies (mainly fishery commissions) established under such conventions. There are also a number of international fishery agreements of a more temporary nature. The United States, for example, is a party to eight international conventions and seven international agreements all dealing with specific fishery resources and areas. The principles under which such arrangements are made vary depending upon the biological characteristics of the resources under exploitation, the stage of development of fisheries in question, the interests and the organization of the fisheries of the nations concerned, etc. The results are usually not entirely satisfactory to any of the parties involved but generally contribute toward minimizing international disputes, avoiding the disruption of major fishing activities, and maintaining yields from the resources concerned at levels higher than they would be without such arrangements.

The above outlined regime for the regulation of high seas fisheries is quite imperfect. The regime has been criticized for its inefficiency and a lack of principles applicable on a world-wide basis. Yet none of the alternative proposals that have been presented so far appear to be workable. Most of these proposals ignore the historical background of fishery development, the diversity of interests among nations, and problems of implementation—in fact most of the practical aspects of international arrangements for high seas fisheries. They also do not pay enough attention to the fact that some of the existing arrangements go far beyond the question of conservation (for maximizing total physical yield) and include a variety of measures to cope with the question of who gets what. If the parties to a convention can agree, almost anything is possible.

I am afraid the existing regime will remain the only practical one for some years to come. Any serious attempt to change it drastically would perhaps bring about a further reduction of the part of the ocean which could be used by any nation capable of exploiting living resources therein and result in a slow-down of fishery development in many parts of the world.

What the international community should attempt is not to change the basic aspects of the existing regime but to make it more workable and expand its coverage to avoid the depletion of important resources and prevent unnecessary international disputes. The present regime has many weaknesses, but it has a definite advantage in that it is still relatively flexible so that further developments in the exploitation of marine living resources could be accommodated within its framework. In view of the great potential of food resources from the ocean and the possible technical innovations that are not foreseen at present, it would be a mistake to codify too rigidly fishing and related activities in international waters. What looks like a good principle for regulation of international activities might become a serious obstacle to development in the future.

INTERNATIONAL EFFORTS FOR DEVELOPMENT

In the last several years, almost everybody has been talking about ocean exploration. This is understandable, for the ocean covers roughly seventy percent of the earth's surface and we still know very little about this part of our own planet, while billions of dollars are being spent for space exploration. However, if the ocean is to be studied for the benefit of all mankind, as is often advocated, then it is obvious that major efforts must be made to survey the ocean for the development of living resources. Yet if you examine the distribution of funds and efforts in ocean exploration, you will find that not much support is given to fishery surveys and research for this purpose. Of course research of any kind may be carried out for national objectives or simply because it is considered a challenge. But it should be pointed out that the aspects of ocean resource development which are of immediate practical importance to mankind are not receiving enough attention.

The international community should give more serious consideration to the possibility of greatly increasing its efforts in the field of fishery surveys and development. The emphasis of such efforts might be on surveys of under-utilized or unused resources, as well as on the development of methods to exploit them. At the present rate of expansion of fishing activities on a global basis, readily exploitable resources will become more and more difficult to find. In addition to fishery research and survey programs carried out by individual nations, it would be desirable for coordinated international programs of large magnitude to be conducted to cover wide areas of the ocean.

Some people argue against such surveys under international auspices on the grounds that the results would be used mainly by strong fishing nations to exploit resources in waters off the coasts of the developing nations. Although there may be situations in which such an argument appears valid, it is a rather shortsighted view. As mentioned before, the developing nations have generally done rather well in pursuing possibilities of fishery development. The long-term, general trend will be for them to take increasing shares of food production from the ocean. Although the degree to which each of the countries concerned can benefit from an international fishery survey program would vary depending on the status of their fisheries, they would all benefit. The merits of such a program should be considered in this context, and it is unrealistic to expect that all nations concerned should benefit from the program to the same extent or in a similar fashion.

As long as the living resources of the ocean are utilized as common property, all nations have responsibility for making contributions to the development of new resources. It should also be pointed out that one of the common problems of fishery development has been the lack of reliable information on the potentials of the available resources. Many ambitious fishery schemes, for example, have failed due to an unrealistic appraisal of the availability of raw material. The objective analysis and interpretation of the results of an international program would help prevent millions of dollars from being wasted on ill-founded projects.

Although there are bilateral and multilateral programs of wide variety for international assistance in fishery development, fishery projects financed by the United Nations Development Programme (UNDP) and executed mainly by the Food and Agriculture Organization of the United Nations (FAO) now represent the greatest effort from a single source, both in terms of funds and the number of specialists employed. Should the international community decide to boost its efforts to increase food production from the sea, this setup might be used as a catalyst to mobilize the resources of different nations to carry out international programs.

It should be recognized, however, that ocean research is in some ways more difficult than, for example, space research. While most things in space are visible and can be located precisely, almost everything in the sea is not visible. Indirect methods have to be used for most of the observations. The knowledge of the basic characteristics of the ocean, such as topography and physical properties of water, as well as their fluctuations, is still inaccurate. The estimation of the abundance of any living resource is bound to be very rough except for those things we can actually count, such as salmon coming upstream. This means that a greatly increased amount of money must be made available to carry out large-scale and effective regional fishery surveys or development projects.

A project called the Indian Ocean Fishery Survey and Development Program has been proposed by the Indian Ocean Fishery Commission, which was established by FAO with the participation of both developing and developed countries. The proposed program is the first ambitious undertaking to survey the fishery resources in a large area of the ocean through the joint efforts of nations. If such a program can be implemented effectively, it will be a historic event. The possibility of the UNDP supporting the preparatory phase of the program is now under consideration. The proposed program should be clearly distinguished from the International Indian Ocean Expedition conducted a few years ago, which was basically an oceanographic program and did not produce much information useful for fishery development purposes. The Indian Ocean remains a grossly under-exploited and poorly studied region as far as fishery resources are concerned.[6] The whole task, including the preparatory phase and the actual survey and development program, will be a difficult one. It is hoped, however, that this first attempt will not be a complete failure. If nations cannot be persuaded to participate in such an undertaking, the sincerity of the international community about developing ocean resources for the benefit of mankind will be subject to serious doubt.

FRANK L. LAQUE

9 Deep-Ocean Mining: Prospects and Anticipated Short-Term Benefits

The expectation of considerable revenue from the exploitation of ocean mineral resources has generated a great deal of discussion during the past few years. Until now these discussions have concentrated heavily on how exploration and exploitation of the anticipated resources should be made subject to some form of international regulation and on how the revenues from such exploitation should be applied for the benefit of mankind. Underlying this discourse is the concept that deep-ocean mineral resources represent "a common heritage" and that, therefore, the wealth derived from their exploitation should be held in trust by the international community and applied for the common good. There are some, also, who feel that the anticipated riches from ocean exploitation should go toward redressing the imbalance between the developed and the developing nations of the world.

Those holding these views envision two prospects for the future. One suggested possibility is that a wild international scramble will take place among the highly developed nations to dominate the exploitation of undersea resources. International tensions would consequently be aggravated and the advanced nations would become even more prosperous in relation to the developing ones. The other suggestion is that an enlightened international social conscience will result in a general recognition that the substantial (often called tremendous) new resources in the ocean can

FRANK L. LAQUE is Vice President (retired), International Nickel Company, and President, International Organization for Standardization.

provide mankind with a splendid opportunity. Imbued with a generous new spirit, men may seize the chance to organize exploitation so as to eliminate any possibility of increased international tension and may then distribute the derived wealth for the maximum benefit of mankind, with special concern for developing nations.

The purpose of this paper is to evaluate the prospects for exploiting deep-ocean metals and to examine the possibility of achieving the proposed international goals as a result of this exploitation. This study will deal only with metals—those that are sometimes called "hard minerals," as distinguished from such other minerals as petroleum, natural gas, sulfur, and phosphorites. Sand, gravel, diamonds, precious coral, and the like, will be excluded.

DEEP-OCEAN VS. COASTAL DEPOSITS

Since the ocean mineral resources of present concern are those that may become subject to some sort of international regime, the probable boundaries of this international regime will determine the nature and location of the resources involved. Presumably the Geneva Convention on the Continental Shelf of 1958 defined the limits of jurisdiction of coastal nations over the resources of the seabed. The limit established by a water depth of two hundred meters was made much more imprecise by extending that limit to any depth capable of exploitation, subject to further limitation by a criterion of adjacency to the coastal state claiming jurisdiction. Experience has shown that each country is likely to have its own interpretation of the provisions of the Convention in extending its limits of jurisdiction beyond the two-hundred-meter depth line, and the technology for drilling oil wells has already advanced well beyond the two-hundred-meter isobath.

We may realistically assume that nations will continue to assert and defend claims to national jurisdiction over seabed resources beyond the two-hundred-meter depth limit. Indeed, both geology and the record of discussions leading up to the Convention of 1958 have been claimed to provide a substantial basis for interpreting "adjacency" as extending to the outer margin of the submerged continental land mass—to the area where this submerged land mass meets the different geological structure of the abyssal ocean bottom at the edge of the continental slope.[1]

Any offshore mining operation undertaken beyond the present depth limits but covered by the adjacency criterion of the Geneva Convention would, therefore, be regulated under national jurisdiction. For all practical purposes, exploitation amenable to international control would thus take place in relatively deep water, at depths probably in excess of twenty-five hundred meters. The first task, therefore, must be to determine what minerals exist at this depth and which ones are possibly recoverable.

Mining operations resembling those on shore and involving the sinking of underground shafts for the excavation of mineralized veins or zones would be difficult and expensive even in the relatively shallow waters of the continental shelf. Although Carl F. Austin asserted in a recent paper the eventual technical feasibility of such operations,[2] they are quite unlikely to be attempted in the deep ocean in the foreseeable future. Furthermore, as Preston Cloud has pointed out, "Modern theory of sea-floor spreading implies that beneath a thin veneer of later sediments the ocean basins are generally floored with relatively young and sparsely mineralized basaltic rock,"[3] and Harold James has reached a similar conclusion.[4] It would be safe to say, therefore, that underground mining in deep international waters is such a remote possibility that it need not concern us at present.

Placer-like deposits of gold, silver, platinum, tin, and diamonds eroded from onshore mineral deposits and carried into the ocean by streams cannot be expected to extend beyond the limits of national jurisdiction, which will encompass the lowest sea level in geologic times when a large fraction of the earth's water was tied up in ice on land. Many areas near shore now covered by water were then exposed, and river beds, possibly containing placer deposits of metals, were then on land. The same is true of mineral-rich beach sands containing valuable concentrations of titanium, zirconium, and iron.

With regard to elements dissolved in seawater, commercial exploitation has been limited to magnesium, bromine, and common salt. Since these are readily available from coastal waters, they are of little interest in terms of the exploitation of deep-sea minerals. The concentration of other metals dissolved in seawater is so low that there is practically no chance of their profitable exploitation; tremendous volumes of water would have to be processed to recover any significant amount. For example, the treatment of five hundred million gallons would yield only about ten pounds of nickel, and concentrations of other metals of interest are of the same order of magnitude.[5]

Metal-enriched muds associated with hydrothermal activities, like those that have been explored in the Red Sea [6] cannot be expected to be extensive enough, sufficiently rich, or widespread enough to warrant our considering them of immediate significance in the total exploitation of metals from the deep ocean. The same applies to consolidated-vein or lode deposits that might occur at relatively shallow depths on ocean ridges.[7]

Since the effect on deep-sea exploitation in the near future of some of the metals and operations mentioned above is as yet uncertain, this paper will be confined to a discussion of the primary hard mineral resources in the deep oceans. These are manganese and the associated metals found in nodules lying on the ocean floor. As V. E. McKelvey, J. I. Tracy, G. E. Stoertz, and J. G. Vedder have pointed out, "The manganese nodules, in fact, are the only likely potential resource over much of the large ocean basins. . . ." [8]

The existence of manganese nodules on the deep-ocean floor has been known since the famous Challenger Expedition of 1873–1876.[9] Since then numerous other explorations have provided evidence of a wide distribution of manganese nodules of varying composition and potential value. V. E. McKelvey and F. Wang of the U.S. Geological Survey [10] have recently published maps showing locations from which nodules have been recovered in exploratory surveys. So far, only a very small fraction of the total ocean-bottom area has been surveyed, but explorations to date have shown that nodules of attractive metal content are most likely to be found at depths in excess of twelve thousand feet (thirty-six hundred meters).

Despite the limited extent of current exploration, considerable evidence [11] indicates the presence of manganese nodules over broad ocean areas. This conclusion would be borne out by the general uniformity of ocean-water sources of the nodule constituents above large areas of the ocean bottoms.

As yet, exploration has been insufficient to establish firmly the existence of specific areas covered with nodules of exceptionally high, valuable metal content or to delineate the boundaries of any such areas. Such "hot spots," if found, would constitute unusually desirable concentrations for exploitation and would spark a demand for exclusive concessions in contrast to a generally recognized right of anyone to exploit the nodules of common value distributed over broad areas of the ocean floor, all equally attractive.

Should "hot spots" be discovered, especially if they prove to be relatively rare, international mechanisms would be needed for granting and policing concessions. Beyond the rarity and richness of possible concentrations, other factors would tend to make some locations more attractive than others. Desire for tenure of defined areas would be influenced by such features as: their proximity to potential markets for the metals recovered; their nearness to land bases for refining plants, as a consideration of costs; meteorological and sea conditions; the depth of water in which recovery operations would have to be undertaken; the topography and the soil mechanics of the bottom; and the political stability, overall business climate, and other conditions in adjacent coastal nations where supplementary land-based operations would take place.

The commercial value of nodules will amount to the difference between the market value of the extractable metals and the cost of finding and recovering the nodules, transporting them to refining plants, extracting the metals in marketable forms, and marketing them.

The total income from all nodule-exploitation operations and the total area of the ocean bottom involved will be determined primarily by the composition of the nodules and their concentration in terms of pounds per square foot of ocean bottom. Concentrations as high as seven pounds per square foot have been estimated from photographs. A more reasonable and more conservative estimate for purposes of discussion would be two pounds per square foot, equivalent to 27,878 tons per square mile.

The composition of nodules can be expected to vary through wide limits. The principal constituents will be manganese and iron and the most valuable nickel, copper, and cobalt, always in much smaller concentrations. On the basis of the limited number of samples available for analysis, we could define a representative composition of Pacific Ocean nodules of possible commercial interest:

Manganese:	25	percent
Nickel:	1	percent
Copper:	.75	percent
Cobalt:	.25	percent

A typical Atlantic Ocean nodule contains these elements in a different ratio:

Manganese:	16	percent
Nickel:	.42	percent
Copper:	.20	percent
Cobalt:	.31	percent

Because of the inadequate number of available analyses, calculations showing the average metal content of nodules would have limited significance. Furthermore, the iron content of nodules is too low—generally under twenty percent—for it to be assigned any value in an appraisal of the potential market value of the metals in nodules.

Although the metal content of nodules varies over different areas of the ocean bottom and predictions can be made only in broad, general terms, sampling to date indicates that the greatest concentration of valuable metals is in the Pacific Ocean rather than in the Atlantic and that the most extensive nodule-recovery operations will therefore probably take place there. Unfortunately, nodules of potentially attractive commercial value seem most likely to be found at very great depths of water, from about twelve thousand to eighteen thousand feet (thirty-six hundred to fifty-four hundred meters).

The content of associated metals in some nodules will be high enough to make the manganese unsuitable for its major fields of application unless the associated metals are removed. At the same time, the amount of these metals may be so small and their value so much less than the cost of refining the manganese for their removal that the nodules will be economically unattractive.

As Table I indicates, there is a disparity between the ratio of metals in nodules and the ratio of world demand for them. We can appreciate the dramatic implications of this disparity by noting that if the world's current need for copper were to be supplied completely from the exploitation of nodules, there would be made available at the same time nearly twenty-five times as much manganese, fifteen times as much nickel, and a hundred and thirteen times as much cobalt as the market could absorb. Probably the most important conclusion to be drawn from Table I is that expected revenue from the exploitation of nodules cannot be calculated simply by adding up the value of the individual metals per ton of nodules. The assumption that there will be a market at current prices for all the metals in the nodules is unwarranted.

Some discussions on the effect of recovering metals from nodules have concentrated on the impact that metals thus derived might have in lowering market prices. Such calculations and predictions have failed to take into account the more important question stemming from the data in Table I—the extent to which the metals might be able to find a market at any price. In the light of present knowledge, there is no reason to expect that individual metals can be recovered from nodules at a cost less than that of mining land-based deposits. From this it follows that the exploitation of nodules will be economically attractive only if a market can be found for more than one of the metals present, and it seems unlikely that the recovery of manganese from nodules will be economically attractive.[12]

Depending on the process used, the form in which the manganese is made available, and the cost of shipping it to market, the manganese in nodules might have some value, but the price of manganese would probably drop as a result of adding nodules to existing sources of supply. On the other hand, the manganese might be discarded in the refining process as "rock." In that case, no realizable value would attach to the manganese content of nodules.

The economic attractiveness of manganese in nodules could be increased by a successful effort to develop large new uses for this metal. If the goal is to increase the total value of nodules, however, such new uses should not compete with those of the other metals associated with manganese in nodules. Efforts to develop a market for manganese that is independent of steel production have been unimpressive in the past, to say the least. If the exploitation of nodules for manganese is to be made commercially attractive, more research in this area is required.

The same considerations apply to cobalt. Although the effort to develop new uses for cobalt has already been considerable, more is needed to improve the future of nodule exploitation.

Table II shows the tonnage of nodules of the composition chosen as a model that would have to be harvested, the areas of ocean bottom that would have to be exploited on the basis of two pounds of nodules per square foot, and the fraction of the total ocean bottom that would have to be worked to produce metals from nodules to equal world production from land sources in 1967.

From the data in Table I, we could expect that a nodule-exploitation operation would encounter the least difficulty if it were aimed at satisfying a major share of the world demand for cobalt and that the early stages of nodule exploitation might well be geared therefore to the world's need for cobalt. Based on the data in Table II, the maximum limit of exploitation would be about 6,500,000 tons of nodules per year. Exploitation at this level would yield approximately thirty-three million pounds of cobalt, four million tons of manganese ore, a hundred and thirty-two million pounds of nickel, and a hundred million pounds of copper.

It would not be realistic to assume that over twenty percent of the world market for manganese could be displaced immediately to accommodate manganese from nodules, and it remains questionable that the treatment of nodules for recovery of manganese can be made economically attractive. We can reasonably assume, therefore, that the real metal value of nodules would lie in their nickel, copper, and cobalt content. An estimated gross revenue of about $285 million would result from meeting the 1967 world production of cobalt. This gross would be reduced by the costs of recovery, refining, and so forth, and a net revenue before taxes of sixty million dollars would be an optimistic figure. An assumed international tax rate of fifty percent would yield thirty million dollars for possible distribution to developing nations. Giving value to the manganese would increase the tax revenue only by about ten million dollars.

As Table II indicates, satisfying all the world's need for cobalt in 1967 would have required harvesting nodules from an area of ocean bottom measuring only 236 square miles, which would comprise only 1.7 ten-thousands of one percent of the total ocean area. Even if we went to the unlikely extreme of abandoning all land-based sources of the metals involved and supplied the world's needs for all these metals exclusively from ocean nodules, only about .02 percent of the ocean bottom would require harvesting each year. In other words, one percent of the ocean bottom could be expected to satisfy the world's needs for manganese, nickel, copper, and cobalt for about fifty years, in terms of the demand in 1967.

From the data already available and on the assumption that as much as 1.7 trillion tons of nodules distributed broadly over large areas of ocean bottom may be found in the Pacific alone,[13] it would therefore be reasonable to expect that a minute fraction of the ocean bottom will yield the total world need for the metals involved. If this is the case, an international regime would have to deal with only relatively small areas being exploited simultaneously. The few individual operations necessary to meet the demand need not and would not be likely to interfere with one another.

Several factors will influence the extent of nodule exploitation in the future. First and foremost will be the availability of land-based ores of equal or superior commercial attractiveness. The number of years of supply represented by these reserves, taking the 1967 rates of production as the standard, are summarized in Table IV, but these estimates are sure to be extended by the discovery of new ore bodies on land and by the development of techniques for recovering and treating lower grade ores. It can be concluded from these tabulations and from the studies of V. E.

McKelvey[14] that the exploitation of deep-sea nodules will not amount to a desperate attempt to compensate for the exhaustion of land-based sources of metal at any time in the near future.

If the cost of recovering metals from deep-sea nodules were found to be less than the cost of exploiting from land-based ores, there would naturally be a strong incentive to abandon land-based sources. There is no present evidence, however, that recovery of metals from nodules will be more profitable than land-based exploitation. P. E. Sorensen and W. J. Mead[15] concluded, on the contrary, that at the present time the exploitation of nodules for their metal content cannot be expected to be profitable even if credit is allowed for the manganese content. They based their conclusion on estimates of the capital cost of recovery equipment (dredges) and transportation, together with refining costs. While this conclusion might be unduly pessimistic, the commercial advantage of exploiting deep-sea nodules remains to be demonstrated.[16]

If and when the exploitation of metals from nodules becomes commercially attractive, a limitation on the scale of operations may need to be imposed by some international agency. Regulations may restrict the volume of production to conserve resources or to minimize interference with profitable markets for metals mined on land.

Restraint may also result from the unwillingness of land-based producers to abandon mines and processing facilities in which they have a large capital investment. We can expect some recalcitrance from these producers since they would be faced with the simultaneous necessity of raising new capital for the exploitation of ocean nodules. The capital requirements for handling the very large tonnages of nodules involved could easily approach many billions of dollars for the total shown in Table II.

Restrictions may also result from national and international restraints on potential exploiters for the purpose of protecting national sources of tax revenue and preventing unemployment in land-based mining industries. Countries currently depending for their prosperity on the exploitation of land-based ores might be expected to exert pressure on international control agencies to restrain deep-sea exploitation.

In some instances the exploitation of metals from nodules may be encouraged or expedited for strategic reasons by nations wishing to end their dependence on remote sources under the control of possibly unfriendly nations or to eliminate the hazards of long-distance transport. Their inclination to do so will diminish, however, if the cost of metals recovered from the sea is substantially higher than that of metals obtainable on land.

THE ANTICIPATED BENEFITS FROM DEEP-OCEAN MINING

In the light of the uncertain future of deep-ocean mineral exploitation and its yet-to-be-established commercial value, the prospects for using revenue from this source to help developing nations seem poor.

In terms of prosperity, the nations of the world range from affluence to poverty on a sliding scale. Because the variations are gradual, an agency charged with distributing tax revenue from deep-ocean mining, even if it were substantial, would have difficulty deciding which developing nations were entitled to a share and how the total should be allocated among them.

The gross national product of a country is a reasonable measure of its prosperity. For purposes of this discussion, it is assumed that developing countries, as candidates for revenues from deep-ocean mining, would be found in Latin America, South Asia, the Near East, the Far East (except Japan), Africa (except South Africa), and Oceania (except Australia and New Zealand). The total gross national product for these areas in 1967 amounted to 12.6 percent of the world G.N.P., or $291,254,000,000.

Figures for the proportion of the world gross national product represented by the value in 1967 of manganese, copper, nickel, and cobalt, the valuable constituents of nodules, are given in Table V.

It can be calculated that the value of world production of manganese, copper, nickel, and cobalt in 1967 represented only .28 percent of the total gross national product. The distributable revenue from taxation, ten percent of the total value, would be about .028 percent of the world G.N.P. It may be noted that the total world production of these metals in 1967 had only about one-half the value of the world catch of fish in that year.

In addition, a substantial portion of the world's production of manganese (23.1 percent), copper (41.7 percent), and cobalt (89 percent) comes from developing countries. While most of the world's nickel now comes from Canada, New Caledonia stands second in nickel production, and new nickel projects are in various stages of exploration and development in New Caledonia, Guatemala, the Dominican Republic, Indonesia, the Philippines, and the Solomon Islands. [Details of the value of metals production of developing countries in 1967 were presented in a similar paper by the author before the Marine Technology Society in Washington, D.C. in June 1970 and to be published by this Society.]

If, as would be the case, only the revenue from taxes on the profits deriving from the exploitation of deep-ocean metals is available for adjusting the relative prosperity of developed and developing nations, this amount would be about ten percent of the total market value of the metals and would represent only a little more than .025 percent of the world gross national product and only about .2 percent of the G.N.P. in 1967 of the developing nations. On a per-capita basis, this would come to forty-one cents a head if it were divided equally among the 1,594.9 million people in the developing countries.

It should be evident, therefore, that even in the unlikely event that the deep-ocean bottom replaced all land sources of manganese, copper, nickel, and cobalt, the assignable revenue from the exploitation of these deep-ocean metals could have little impact on efforts to close the current gap between developed and developing nations. Furthermore, substituting ocean for land sources of these metals would tend to detract from, rather than to advance, the prosperity of those developing nations with large deposits of metal-bearing ores.

CONCLUSIONS

Most of the current activity in the recovery of metals from deep-ocean nodules can be characterized as an examination of the technical and economic feasibility of various conceptual approaches. Some of these may lead to preliminary or pilot-scale projects that will precede full-scale commercial operations. No such operations are taking place at present,[17] and the aim now is to provide a basis for future decision whenever new sources of ore may be needed. Such an eventuality may occur when per-capita consumption of metals in developing countries approaches the present level in the advanced nations.

While the future of deep-ocean mining cannot be predicted with precision, it is safe to draw a few general conclusions:

1. There will be no commercial-scale exploitation of deep-ocean nodules for several years—probably not before 1980 to 1985.
2. There is a need for an international program of ocean exploration that could be part of the International Decade of Ocean Exploration proposed in 1968 by the former American President Lyndon B. Johnson to confirm the extent, the distribution, and the possible value of metals in deep-ocean nodules.
3. Since exploitation operations in the foreseeable future will probably be few in number and conducted on a small scale, any international control agency or mechanism should place emphasis on providing a regulatory environment, either national or international, that will provide incentives for risky exploitation and will not place the operations under undue restraint. Unnecessary restrictions can result from efforts to deal with unknown situations and circumstances that may never be encountered.
4. While appropriate international regulations will be needed in the future, details should not be worked out before the facts are in hand. International laws or regulations aimed at a codification of practice should logically await reasonably precise knowledge of the practice that is to be codified.
5. Since we currently do not know how much revenue for "the benefit of mankind" can be expected from the exploitation of nodules and since the amount will probably be small for the foreseeable future, the prime international emphasis should be on encouraging exploration and preliminary exploitation rather than on the disposition of revenue. Whatever revenue

TABLE I: Primary metal to be recovered from nodules to extent of total world* production in 1967

Metal	1967 world production	Pounds per ton of nodules[1]	Percentage of 1967 world production of associated metals that would be made available simultaneously: Manganese	Copper	Nickel	Cobalt
Manganese	18,650,000 short tons ore	—	100(%)	4(%)	59(%)	453(%)
Copper	11,184,377,000 pounds	15	2,502	100	1,479	11,335
Nickel	1,007,943,000 pounds	20	169	8	100	766
Cobalt	32,890,000 pounds	5	22	.9	13	100

* Mainland China not included.
[1] Based on nodules containing 25 percent manganese, 1 percent nickel, .75 percent copper, and .25 percent cobalt.

TABLE II: Tons of nodules and bottom areas to be harvested each year to yield metals at the 1967 level of production from land sources

Metal	1967 world production	Pounds per ton of nodules[1]	Short tons of nodules required[2]	Area to be harvested sq. miles	Fraction of total deep ocean bottom area[4]
Manganese	18,650,000 short tons ore	—	29,800,000[3]	1,069	0.0008(%)
Copper	11,184,377,000 pounds	15	745,625,100	26,746	0.0192
Nickel	1,007,943,000 pounds	20	50,397,150	1,808	0.0013
Cobalt	32,890,000 pounds	5	6,578,000	236	0.00017

[1] Based on nodules containing 25 percent manganese, 1 percent nickel, .75 percent copper, and .25 percent cobalt.
[2] Based on nodule density of 2 lbs. per sq. ft. of ocean bottom or 27,878 tons per sq. mile.
[3] Increase due to lower manganese content of nodules (25 percent) as compared with 40 percent in land-based ores.
[4] Estimated to be 139.5 million sq. mi. (361×10^6 sq. km.).

does accrue from deep-ocean mining will probably have no significant effect on the absolute or relative prosperity of the recipients and may well have a greater effect on the distribution of prosperity among developing nations than on the comparative position of the developing and developed countries.

6. Developing nations should not be encouraged to expect that the exploitation of deep-ocean metals will provide a major component of the funds they need for future development.

TABLE IV: Apparent years supply of metals in known land ore reserves at 1967 rate of production *

Metal	*Indicated years supply***
Manganese	98
Copper	55
Nickel	148
Cobalt	146

* From Tables 3, 4, 5, and 6.
** Assuming no additions to reserves from new discoveries or otherwise.

TABLE V: Value of world production of nodule metals in 1967

Metal	*Total Production*	*Market Price*	*Value*
Manganese	18,650,000 short tons[1] ore	$25.68 per ton[2] ore	$ 478,932,000
Copper	11,184,377,000 pounds[3]	.45 per pound[4]	5,032,970,000
Nickel	1,007,943,000 pounds[3]	.90 per pound[5]	907,149,000
Cobalt	32,890,000 pounds[1]	$ 1.85 per pound[1]	60,846,000
Total Value			$6,479,897,000

[1] U.S. Bureau of Mines.
[2] Based on 40 percent Mn content ore @ 72¢ per unit or $25.68 per ton.
[3] *Metallgesellschaft* Statistics.
[4] Estimated composite price.
[5] Average price per year.

EDWARD WENK, JR.

10 Toward Enhanced Management of Maritime Technology

While marine charts and terrestrial maps representing global geography are inherently complementary, they have had one characteristic difference. Navigational charts delineate primarily the irregular boundary between water and land, together with details of submarine topography, tides, currents, and prevailing winds. Terrestrial maps portray similar topographic detail, with emphasis on rivers, mountains, and cultural features such as roads and cities. They differ from marine charts, however, in that they contain additional cartographic information on the boundaries of private property and demark areas of political sovereignty. This distinction between the two types of maps reflects their historically different uses. Nautical charts facilitated safe and speedy transit between ports but were never intended to convey legal subdivisions; recording proprietorship, cultural settlement, and economic development on ocean charts was considered not only irrelevant but unthinkable.

This attitude has persisted until recently. Today, however, the sea has assumed significance beyond its former primary role as a medium for transport. Reflecting this extension in values after years of congressional probing, a new national policy was adopted in the United States in June, 1966. The aim of this policy is to intensify study and use of the sea to enhance human life through

EDWARD WENK, JR. is Professor of Engineering and Public Affairs at the University of Washington in Seattle. He was formerly the Executive Secretary of the National Council on Marine Resources and Engineering Development, Washington, D.C.

economic and social betterment and to promote world peace and understanding. With the world population outpacing its food supply, with industrial requirements for energy and minerals growing faster than population, and with increasing concentrations of waste being unwittingly injected into the marine environment, it had become clear that neither the problems ahead nor the solutions proposed for them could terminate at the water's edge. The policy enunciated in the Marine Resources and Engineering Development Act of 1966,[1] reinforced by subsequent Presidential statements and elaborated in private conferences, communicated both a new determination to relate the sea to the affairs of man and a desire by the United States not to "go it alone."

The international community was encouraged to collaborate in employing the seas for the benefit of all mankind. Corresponding U.S. initiatives at the United Nations General Assembly in the fall of 1966 led to a resolution [2] reflecting the realization that a multinational approach to peaceful uses of the sea was not only desirable but necessary: scientific study of the sea is inherently international in character; deep-ocean resources are the common property of all nations; and the need now is to minimize potential conflict as more and more nations project their interests seaward. Ambassador Pardo of Malta aroused interest at the United Nations with his dramatic proposals for using seabed resources to alleviate the economic peril of the less developed lands and of the United Nations itself.

Many nations began their own inquiry into their stake in the oceans; by now virtually all of the hundred and twelve states bordering the sea, and the twenty-nine that do not, are considering their national interests at a policy level, not just at a scientific level. Whetted by projections, based on enthusiasm as well as on fact, that vast wealth was to be derived from the sea, many responded to the stimulus by mapping the marine equivalent of a terrestrial frontier. They drafted boundaries for possible seabed territorial claims. The cautions expressed by President Johnson in July, 1966, not to "allow prospects of rich harvest and mineral wealth to create a new form of colonial competition—and a race to grab and to hold lands under the high seas" initially went unheeded by nations and special interests alike.

Collectively, the countries of the world have thus begun a debate on three major issues concerning seventy percent of the earth's surface: Who owns the seabed? Beyond boundaries of natural sovereignty, how will exploitation be controlled? How will benefits be distributed?

A NEW APPROACH

It is my thesis that to extract the sought-for benefits from the sea we need a fresh approach flowing from concepts of use and from the relationships among the institutions involved as well as from legal principles and ideology. To aid this mode of analysis, a second category of maritime charts will be needed. Instead of hypothetical boundaries of national sovereignty, these charts would follow the principles used in maps delineating the economic geography of occupied land areas. They would be concerned with the present and future use of the seas.

Two questions arise: First, what intelligence should be portrayed on these maps? And to answer that question, the second—what is their ultimate purpose? I would propose, as an orientation for such charts, a rationale leading to the enhanced management of maritime technology.

To emphasize maritime technology is to reflect the significance of the recently acquired engineering knowledge that is making these resources accessible and exploitable. Our heightened interest in marine resources is a direct consequence of that new knowledge. We now have the technical muscle to accomplish those feats on, in, or under the sea we have long wished for but have always been denied because of the strenuous maritime environment. Maritime technology has generated the use of new and sophisticated tools for exploration—spacecraft, buoy networks, and submarines—as well as tools for the exploitation of marine resources. Technology is properly thought of as concerned with means, but its management involves the notion of ends as well.

The concept of managing technology is a relatively new abstraction and has entered public discussion only recently. This concept has taken hold as technology, managed ineffectively by human institutions, has been seen to inflict damage on our environment. The application of a concept such as this even to familiar enterprises on land is tentative and untried. Application to the marine environment is all the more novel. The object of this paper is to open discussion on the validity and implications of the concept in meeting the conflicts emerging over use of the sea and the seabed and in illuminating opportunities to lay a sound, if unprecedented, basis for future international development. The paper also establishes functions for proposed new supranational machinery in order to assure enhanced management of maritime technology.

First, a definition. The word "technology" is meant to convey the complex process by which a technique is successfully applied to achieve a selected purpose. This definition implies that the technique itself is specialized and refined for the intended purpose—a maritime one, in the context of this discussion. In contemporary terms, techniques are based more on scientific and engineering research than on empirical experience or craftsmanship.

If it is to translate specialized knowledge into effective accomplishment, the technique must be afforded an appropriate institutional vehicle. This vehicle—or enterprise—must have the capacity for decision-making, especially with regard to risk, and the means for raising capital, mobilizing specialized manpower, and articulating with both the marketplace and other enterprises that surround and affect it. It must provide a platform for leadership.

Technology requires that such an institution have the capacity to cope with change and to innovate. This requirement, in turn, implies the ability to collect and analyze information and to undertake "pre-crisis" as well as "post-crisis" studies. The enterprise must be able to generate new knowledge, that is to say, to perform the research necessary to extend and refine the techniques on which the enterprise was initially founded. It may even spawn subsidiary enterprises whose only goals would be to extend knowledge through scientific research, geographical exploration, or engineering development.

Increasingly significant to contemporary technologies is development of an harmonious relationship among institutional components drawn from four familiar groupings: national governments; international organizations; corporate enterprises; and academic institutions. Historically, the requirements for technology have been generated by both public bodies and private entrepreneurs. Military weaponry, space exploration, and nuclear energy are examples of areas where technology was initially government-sponsored and where the other groups came to be involved later. The production of consumer goods is an obvious example of the second.

The international institution may well play a special role in attempts to deal with maritime technology.

The management of maritime technology, or any other technology for that matter, involves stimulation, control, administration, coördination, and regulation of all components of the technology, so optimized that the aggregate activity may best serve a particular purpose. It has recently been recognized, however, that such management requires an "early warning system" and a corresponding discipline if the newly generated technology is not to induce unwanted side effects inadvertently, through narrow application. It is urgent to recognize this precept of management in the present era of transition from a time when man utilized technology to protect himself from the environment to one in which technology is needed to protect the environment from man.

Major elements of maritime technology may be directed toward exploitation of marine resources so that the management of maritime technology may be considered to subsume the management of resources. Thus, technology management goes beyond resource management in its concern for the techniques and tools of exploration and exploitation, as well as for the development and conservation of the raw materials themselves. In particular, the concept requires a systems approach that takes into account the interactions that occur during exploitation among the ostensibly separate maritime resources, their different uses, and the activities of the various institutional groups involved in technological development. All potential uses of the sea must be considered rather than merely the exploitation of fish and non-living resources: waste disposal, mercantile commerce; peacekeeping; recreational development; scientific research; and conservation. To seek harmony among these uses would be the object of planning, so that no single-purpose use may unwittingly preëmpt options for the future.

The concept also reflects the premise that sources of capital and instruments of research, discovery, extraction, exploitation, and marketing are as important in developing and distributing benefits from the sea as ownership of the undeveloped resources themselves. In this approach, questions regarding the richness and the distribution of living and non-living marine resources have only subordinate interest; the major concern is to identify and resolve collectively the more fundamental question of what we want these resources for.

FUNDAMENTAL GOALS FOR THE SEAS TO SERVE MANKIND

To manage means to manage with a purpose. If maritime technology had to satisfy only one purpose, the selection of the best management alternative would be much simpler than it is. Discussions at the United Nations have revealed, in the main, three broad purposes that are widely although not universally accepted: preserving world order; maintaining the quality of the environment; and accelerating nutritional and economic health among less developed nations in order to reduce the continuing disparity between them and the technologically advanced nations.[3] These broad goals are not mutually exclusive but they can be both implicitly and explicitly in conflict. Moreover, short-run objectives may not easily be reconciled with long-term goals.

The first step toward enhanced maritime management thus must be the quest for a common purpose, including an effort to discover what is socially desirable, so that we will not simply yield to the seduction of what is technically feasible. The implementation of that step raises questions about the adequacy of our present machinery for universal balloting on an issue like the management of worldwide technology since the constituency includes both national governments and institutions concerned with the use of the sea.

ANALYTICAL CAPABILITIES FOR MARITIME PLANNING

To penetrate this management concept further, it will be necessary to draft the new charts proposed earlier that would portray the economic geography of the oceans. Such charts would, of course, delineate resources, indicating both their presence and their accessibility. Gaps and uncertainties in these charts would immediately suggest targets for geographical exploration. To provide further clues for exploration priorities, another set of charts would be based on the economics of resource recovery, especially in its relation to the exploitation of corresponding terrestrial resources, showing the proximity of marine resources to processing capabilities, energy sources, points of transshipment, and markets. To these resource maps must be added still others not related directly to resources but showing trade routes, recreational sites, tourist traffic, sources of pollution, and ocean circulation patterns that carry pollutants far from their original source. Finally, maps should delineate natural areas of special beauty or sensitive ecology deserving of special protection.

Composite maps of this type should tell us who is involved in what use, in terms of capital investment and profit sharing. Portraying national jurisdictions is a dimension that will lend both complexity and fragmentation to such maps. In addition, if we are to comprehend the full picture, the dynamics of power and influence reflecting the relationships between individual governments and users must be clear, that we may go beyond a simple taxonomy of participants to understand the actual dialogue. It is naive, for example, to equate power with national entities alone. Governments should be identified in terms of their respective roles in providing low-interest capital or subsidies to their nationals, sharing risks, limiting liability, sponsoring exploration and research, regulating pollution, and so forth.

With such charts, it would become evident that the tapestry of use and the nature of the environment—including circulation of sea water, pollutants, and fish—in no way corresponds to projected political boundaries. The actual or potential conflict among the various uses would also be manifest. Shipping lanes simply cannot pass through concentrations of oil platforms, and industrial waste cannot be dumped on oyster beds. Conflicts will naturally be intensified in the relatively shallow water near coastlines, but engineering potential can extend the locus of activity farther to sea. In the light of the conflict between oil extraction and conservation, industrial enterprises may well be forced out to sea.

THE RATIONALE FOR GLOBAL PLANNING

It may be argued that a similar complexity exists on land and that no such detailed land maps have been constructed to deal with issues there. The term "maps" as used here is a rhetorical device referring to the analytical tools necessary for dealing with the ocean.

The sea is, first of all, a vast homogeneous area beyond the reach of national sovereignty (wherever national boundaries may eventually be drawn) and thus is fundamentally a simpler problem than the heterogeneous land masses.

Second, the oceans are now at a sufficiently low relative level of activity and rivalry that many more options are available for its future development than exist for terrestrial activities. Land development has historically been dominated by the concept of private ownership and political subdivision.

Third, we still have the option of keeping the management of ocean technology simple. Many problems in modern society arise from the sheer complexity of its institutions, and, as with biological systems, excessive complexity may be a threat to survival. Coordinating so many different elements with conflicting purposes may fatally overstress the system. Keeping institutional relationships simple could itself be a goal in managing maritime technology.

Society has increasingly failed to manage technology successfully on land. Many of the problems we now face are second- or third-order consequences of a technology adopted for a particular goal. Whereas that goal may have been successfully met, its realization has often been accompanied by serious and unwanted effects that now require emergency correction. More deliberate advance assessment of technological side-effects and better management could have headed off the costs we now face in cleaning up polluted air, water, and land.

Perceptive management becomes all the more imperative in the face of the projected demands that will be made on the ocean during the three short decades left in this century. The anticipated three- to six-fold increase in fishing, oil recovery, ocean transport, waste disposal, and recreational demand, to name but a few activities, makes conflict increasingly probable. The portending conflicts may arise not only between nations but also between users and between institutional groups.

Technology management now has unprecedented means to illuminate the possibly unwanted consequences of any preëmptive development that might conflict with others or impart injury to the natural environment. The advance analysis of which we are now capable can make it possible for us to improve techniques before applying them and thus to minimize the need for subsequent costly or politically difficult corrective measures. The number of oil spills projected for the next thirty years on the basis of recent experience suggests that no coastline will remain unstained. The accuracy with which fish concentrations may be spotted by spacecraft or other devices in the foreseeable future could quickly lead to extermination of species. Setting boundaries will not solve these critical problems.

The advance analysis required will depend on a new form of planning capability whose purpose will be to examine the long-term consequences of the impact of man and his technology on his environment. Concern for the quality of the environment must go well beyond simple control of pollution; quality is not measured only by absence of pollution; it involves also the achievement of harmony among potentially conflicting uses.

A frequent rhetorical question generated by such problems is, "What would you do if you could start over?" In the case of the oceans, it is still possible to ask this question.

What would we do?

In the first instance, I would suggest that attention be directed as much to the question of collective technology management as to the question of ownership, for the means of exploitation may be more profoundly significant than ownership. Unlimited exploitation of resources solely for economic gain, without improved technology, could be a critical threat to the marine environment.

THE ROLE OF LAW

This discussion leads naturally to the role that international law will play. New or amended conventions to define sovereignty will undoubtedly be warranted since the necessarily large capital investments in ocean exploration and exploitation will be made only under the security of a stable legal regime, and the status of relevant conventions is highly controversial. Discussions of sovereignty have been dominated largely by the issue of narrow versus wide seaward extension of national jurisdiction, with the consequence that goals and the barriers to their achievement have received little attention.

Although the concept of technology management set forth in this paper has many legal ramifications, suffice it to say here that international conventions will comprise only a small part of the "rules of the game" required in the future. If we are to govern the relationships among enterprises now involved at sea, and not just relationships among nations, international regulation should provide for the orderly development of resources, for conditions favorable to investment, for the dedication of returns from common resources to world community purposes, and for an harmonious accommodation of all uses, commercial and otherwise, including the preservation of environmental quality.

INSTITUTIONAL INNOVATION

Assuming that large areas of the ocean and seabed will eventually be regarded as common property subject to universal sovereignty, will the various existing enterprises capable of operating in these areas have the vitality to realize the necessary goals? If not, what evolution in institutions is necessary? Numerous proposals have been made for the creation of an international authority to register claims and distribute proceeds. These suggestions are reminiscent of nineteenth-century practices in developing the land frontier. Are they applicable to the twenty-first century? What lessons have we learned about unmanaged technology? In the absence of any new enterprise, what will be the outcome?

Institutions involved in the management of maritime technology must, if they are to be effective, provide for a great variety of functions. First among these is the development of an international consensus on the social and economic goals. Beyond that initial requirement, enhanced management of technology will demand greater knowledge of the environment and improved techniques for transforming scientific discoveries into practical applications. Evaluations of the unwanted consequences of technology will be essential. Capital must be raised, goals and resources matched, exploitation undertaken, products distributed, and benefits disbursed.

Among the existing international, national, private entrepreneurial, and scientific institutions, a number can meet these functional requirements, to a partial extent at least. The United Nations can establish goals. Perhaps twenty-five nations have the oceanographic capabilities to explore the environment. UNESCO's Intergovernmental Oceanographic Commission, the Food and Agricultural Organization, and the World Meteorological Organization are all capable of coordinating explorational research. Perhaps six nations and an unknown, but probably small, number of private enterprises have the capacity for ocean engineering. Some few national and private sources have capital available for investment, as do some international sources, such as the World Bank, and the World Bank can act as an agent for disbursing benefits.

Even on the unwarranted assumption that each of these organizations or groups fulfills a specific function completely, several elements are conspicuously missing. Nowhere does there exist an international planning body capable of collecting, analyzing, and transferring the information needed for decision-making by many individual nations as well as by groups of nations. Similarly lacking is an instrument for coordinating and integrating the various international components of a maritime technology. Nor is there an entrepreneurial enterprise actively seeking use of the sea for the benefit of all mankind; all seem dedicated, rather, to a piecemeal, narrowly opportunistic approach.

The need for additional international machinery has been recognized almost from the genesis of United Nations discussions on the seabed. A wide range of proposals has included suggestions for a modest territorial registration office, for a body to distribute fiscal return from exploitation, and for regulatory and policing machinery. That further study is essential is reflected in the recent resolution [4] calling on the Secretary-General to prepare a study and submit his report to the Committee on Peaceful Uses of the Seabed, which would then report to the General Assembly during the forthcoming twenty-fifth session. The terms of reference of that resolution are limited to "jurisdiction over peaceful uses of the seabed and ocean floor, and the subsoil thereof, beyond limits of natural jurisdiction," but its scope with regard to the seabed encompasses "power to regulate, coördinate, supervise, and control all activities relating to the exploration and exploitation of their resources for the benefit of mankind as a whole, irrespective of the geographical location of states, taking into account the special interests and needs of the developing countries, whether land-locked or coastal." It is hard to imagine any potential powers omitted; the policy-planning instrumentality proposed in this paper, however, is not intended to be limited to the seabed. From this resolution, it is only a small additional step to devise institutional ways and means of achieving implementation.

Discussion of institutional innovation almost inevitably stirs up resistance among existing bureacuracies whose instinct is to be suspicious of change—always deemed a potential threat to survival. Considering the difficulties that beset any international operation, it would be advantageous to minimize violence to present structures, limiting it to those changes necessary to achieve the stated purposes.

From the preceding discussion, it is clear that some international authority is required to manage the technology for uses of the marine environment. The functions of such an authority would include the control of franchises for exploitation but would go beyond the mere registration of claims. The authority would function as a planning and coördinating body and would depend on coöperative effort from a wide variety of individual commercial, national, and international enterprises to meet its objectives. It would, however, have the option of taking initiatives to meet internationally agreed-upon goals if random initiatives by users leave critical gaps. The authority must be equipped to collect and disseminate information about the marine environment and resources that are beyond national sovereignty and that it would be empowered to manage. It would provide information to all nations individually, thus materially assisting those having limited scientific capabilities of their own. It should also undertake long-range studies of resource potential and make management-oriented analyses to predict the consequences of projected uses. By supplementing national capabilities, it should assure the operation of such data-collection, monitoring, and prediction systems as are needed to understand the marine environment and to predict the impact of man's activities on it. The authority would look to some ongoing body, such as the United Nations Development Programme or the World Bank, to dispense proceeds.

What is proposed here initially is a steering mechanism with independent analytical facilities to aid in international decision-making. Its analyses should be based upon, but should not be dependent upon, facts developed by member nations.

It would assess unmet needs and opportunities that cross the jurisdictional lines of international organizations; recommend priorities for exploration; and advocate the development of new techniques necessary for optimal use of the sea, specifying means of control, in such areas as finding and catching fish, drilling for oil, preventing or containing spills, and disposing of harmful wastes. It would coördinate planning for all martime-oriented international bodies; encourage investments to meet maritime goals, including the reinvestment of proceeds available to it for research, exploration, and engineering development; and develop legal, economic, and technological studies for identifying alternative policies and criteria for the use by those international maritime bodies that have been established by treaty.

The last decade has seen our attitudes toward the sea begin to mature. Our scientific knowledge of the marine environment is increasing as is our ability to put our information to practical use through engineering. A clearer understanding of the benefits to be derived by individual nations has also emerged, but the fragmentation of use, sovereignty, and even of study may still lead to abuse of the marine environment. What seems necessary now is a unifying concept of the sea and a strengthening of our institutions, both national and international, so that we may realize the potential of the oceans. A new international planning and coördinating body dedicated to enhanced management of maritime technology would be a step in that direction.

PART THREE:

THE EMERGING OCEAN REGIME

There is today wide-spread—virtualy unanimous—agreement that the legal framework of an ocean regime must be based on the concept that the seabed and its resources, beyond the limits of national jurisdiction, are the common heritage of mankind. Even the Russians who, in the United Nations and elsewhere, held out hardest and longest against accepting the legal validity of this concept, are beginning to mellow, as evidenced during the recent conference of the Interparliamentary Union (The Hague, September, 1970). There, for the first time, the Soviet delegation accepted the concept.

The Common Heritage of Mankind is a new concept in international law. Its content must therefore be defined in terms of lex ferenda *(emerging law) rather than* lex lata *(existing law). The Pacem in Maribus project has made some contribution toward the clarification of this concept.*

During one of the Pacem in Maribus preparatory conferences (Rhode Island, January, 1970), Ambassador Pardo explained that he used the term "common heritage of mankind" rather than "common property of mankind," not because he "had anything against property, and I don't express any opinion as to the desirability or nondesirability of this ancient institution—but I thought it was not wise to use the word property. . . . Property is a form of power. Property as we have it from the ancient Romans implies the ius utendi et abutendi *(right to use and misuse). Property implies and gives excessive emphasis to just one aspect: resource exploitation and benefit therefrom."*

Pardo proposed that the content of the common heritage be "determined pragmatically in relation to felt international needs." It is not limited by a complex of real or potential resources. "World resources," he pointed out, "should not be conceived in a static sense. New resources are being constantly created by technology." The common heritage of mankind, however, includes also values. *"It includes also scientific research." Thus, if there were a set of ethical and legal rules to be derived from the principle of the common heritage, these would have to be applicable to science and science policy as well.*

Pardo suggested three characteristics of the common heritage of mankind. First of all, there is "the absence of property." The common heritage engenders the right to use *certain property, but not to* own *it. "It implies the* management of property *and the obligation of the international community to transmit this common heritage, including resources and values, in historical terms.* Common heritage implies management. *Management not in the narrow sense of management of resources, but management of all uses." Third, common heritage implies* sharing of benefits. *"Resources are very important, benefits are very important. But this is only a part of the total concept."*

The concept, thus defined, shows certain striking analogies with the concept of social property *as developed under Yugoslav constitutional law. (Social property has nothing in common with the state-owned property in other Communist Nations.) These analogies are elucidated in Jovan Djordjević's chapter on "The Social Property of Mankind." In Yugoslav theory, social property, like the common heritage, is* nonproperty—the absence of property. *It is "organically tied to the concept of management," and it implies the sharing of benefits and profits.*

Beyond this, Djordjević postulates some further attributes of "social property." The adaptation of these to the "common heritage of mankind" may help greatly in clarifying the legal content of this latter and the structural framework in which it must be embodied.

"Social property" is a process which determines a new relationship, not only between people and resources, but between people —between those who, on the basis of the old concept of property, own the "means of production" and whose who don't. In socialist terminology: between the ownership class and the workers. In a social-property system every worker is a manager, every manager is a worker. The workers do not receive a salary from the employer, but they are entitled to their fair share of the social property and the return in profits therefrom. In the world at large, the "worker" would be the "developing nations"; for although they may well be "owning" resources in abundance, these will not profit them: for the "means of production" are science and technology, and these are "owned" by the developed or "manager" nations. In a common-heritage system both natural resources and science and technology would be non-property and appertain to the world community as a whole. This means a basic change in the relationship between developed and developing nations. The latter would no longer receive "aid," but their share of the common heritage.

Social property, Djordjević points out, postulates self-management (which, in the centralized communist state has to be created) just as common heritage postulates sovereignty (of nations) and autonomy (of enterprises)—which, in the world at large, has to be accepted). At the same time, social self-management (Djordjević) implies planning, and a new mechanism of decision-making which "transforms the traditional representational political structure and creates a new synthesis of individual and common interests, of autonomy and unity."

The same will apply to the concept of common heritage: which, by its own "logic" postulates new forms of decision-making, based on new forms of "representation" (neither the one-state-one-vote system, nor the one-man-one-vote system—both reflecting an an-

tiquated and very limited, political philosophy—will do for an organization charged with resource and environment management), a new synthesis of national and common interests, of national and international law.

That such a political theory can do much in determining the structure or machinery of an ocean regime, is illustrated in the appendix to this volume.

This is not to say, by any means, that political theory can solve the economic and political practical difficulties with which we have to grapple in this real world of ours. Not even the most perfect of political theories can do this. These problems must be solved concretely, practically. Thought and action must concur. But if a correct political theory cannot alone create an ocean regime, it is equally true that not even the cleverest, or most powerful pragmatic action can achieve it if sustained by a wrong and antiquated political theory.

The first chapter of Part III presents (in condensed form) what is today probably the most complete theory of social property and common heritage. Francis Christy in his chapter tests the concept as applied to one particular resource: fish. He comes to the conclusion that the traditional concept of "common property" (implying merely "open access") has run its course; that it must be superseded by the concept of "common heritage" which, he concurs, postulated international institutions of a new type. Arnold Künzly, finally, applies the concept of social property and common heritage to science and its management.

Another, rather voluminous, part of the Pacem in Maribus project dealt with the search for institutional models, or rather, for institutional features of existing international organizations that could be adapted to the functions and needs of an ocean regime. Older organizations, such as the International Telecommunications Union, were thus analyzed—this particular organization offers some interesting parallels between the management of wave-lengths, which are nonproperty, and the common heritage of mankind. Much attention was given to the structure of the

International Atomic Energy Agency, the Development Programme, the World Bank, and the International Labor Organization. While something can be learned from each of these, it became quite clear that an international organization established in the seventies cannot be based on the same principles that gave rise to international organizations a quarter of a century ago or even earlier. The new structure must embody change *and it must be able to* project further change. *The ocean regime must be* sui generis—*as is the environment it is to govern.*

From these comparative studies we have chosen (Chapter 4) the one on the European Community which offers some particularly instructive analogies and precedents. The European Community, like the nascent ocean regime, is a "functional" rather than a "territorial" organization. It makes "law" that is neither strictly national nor international: law that is applicable to States, enterprises and even individuals; while decision-making processes remain essentially founded on consensus among Governments rather than on coercion. With all its defects and setbacks, the European Community points in the direction of a part-economic, part-political, functional, nonterritorial organization which may find more wide-spread application in the future.

The third section of Part III presents three political statements on the ocean regime: one by an American, one by a representative of the developing nations of the third world, and one by a European.

JOVAN DJORDJEVIĆ

11 The Social Property of Mankind

THE CONCEPT OF SOCIAL PROPERTY

Social property is an ambivalent and insufficiently defined concept. The term *social* does not mean appertaining to society in the sense of a new abstraction vested in the existing historical circumstances in a single powerful holder of ownership rights, the state. It is therefore a theoretical error to identify social property with state ownership. Social property does not mean collective or group property—the property of enterprises—nor is it a synonym for the collective property of the people or for some form of social ownership or share-holding.

The term *social property* has a negative meaning; it indicates the *negation of the right to ownership* to each and all. It prohibits the power monopoly over the means of production and the produce of labor. No one—neither state nor community nor enterprise, neither the working collective nor the individual—has ownership rights with regard to the social means of production and the product of labor, nor can he dispose of them as of his property, on the basis of power. This social character of property itself implies a *higher degree,* a more real and more direct socialization of working relationships and the division of productive work. It further implies the establishment of workers' self-management,

---

JOVAN DJORDJEVIĆ is one of the main architects of the Yugoslav Constitution of 1963, President of the Constitutional Court of Serbia, Yugoslavia, and Professor of Constitutional Law at the University of Belgrade.

active self-determination of the producer in the process of production and distribution and in the management of other common social affairs.

Social property is a new historical phenomenon, although the concept is sketched in scattered theoretical systems of socialism, most specifically in the works of Marx and Engels. The establishment of state ownership in the socialist countries has diverted interest in the concept of social ownership and discouraged any research in depth. Only in the wake of the recent socio-economic and political structure changes in some of the socialist countries has it become a subject of social consideration and public interest. Here the Yugoslav experience and its social theory has a special place in the establishment and investigation of this new form of socializing ownership and power. Successive forms of ownership until now have represented shifts within the same kind of ownership. They did not negate it or each other. Social ownership, on the contrary, negates the previous ownership. It is a new scientific concept that restores to ownership its original meaning of common appropriation of the objects of labor and of common distribution of the products of that labor on the part of the producers.

This transition from one to another of these fundamentally different relationships presupposes the existence in the society of socialist processes, freedom from domination and from exploitation and manipulation by power. Such social relations are not ready-made and no global class or political change can establish them as such. They form and develop gradually. *Social ownership is a process,* a process which objectively is still in the first phases of its development.

Social ownership is a collection of relationships between people, not only in the micro-organizations of labor but also in the macro-society; not only in the working organization but in the entire socio-economic structure and in the political system; not only in relationship to the appropriation of the products of labor but also to the alienation from those products; not only in self-management but also in relation to power; not only in the working collective but also in the market.

PUBLIC OWNERSHIP

In Western countries the term public ownership is commonly used to designate nationalized means of production or enterprises that belong to what is called the public sector; the term state ownership is then used in a more narrow sense because public ownership includes ownership managed by local self-management (municipal), and is not considered state-owned, particularly in England. Public ownership is for some a more suitable term than state ownership—which is burdened by "statist" or "socialist" connotations—because public ownership can have, and indeed does have, a self-managing status, and it does not alter the prevailing economic and state systems (capitalism and civil democracy). On the other hand, there are writers, among them a significant number of socialists of globally social-democratic orientation, who believe that all these collective forms of ownership belong if not in the category of socialist ownership then certainly to the group of noncapitalist pro-socialist ownership.

STATE OWNERSHIP

The concept and essence of state ownership cannot be separated from the theory and practice of socialist society. In every socialist country state ownership was the form for nationalization of the basic means of production, and in the majority of them it remains the basic ownership form in industry, in communication, and in commerce, and it is partially so in agriculture. Politically and legally it expresses the transfer of ownership rights from individuals (capitalists) to one generalized entity, the society. Naturally, the state is an abstraction, a generality which is representative and symbolic, but in practice it has been almost anthropomorphized since no abstraction can live on its own, but rather people embody it and give it a real character. Otherwise it is an abstraction remote from physical reality. The state and ownership are very practical abstractions.

State ownership is a legal category which negates the pluralism of ownership rights of individual capitalists but retains individual ownership rights in the form of public rights. Ownership monopoly, it is true, is now in the hands of the state (that is, the groups which designate it), but the essence of the classic class-ownership is not altered thereby. In its best-known—although not unalterable—form, state ownership still expresses relations of command and subordination, a division between those who give and those who take and execute orders in the process of labor, still based on the same relations of hiring and firing. On this basis the state has a new ownership power. As the holder of ownership, the state disposes of both the labor of the producers and the products of that labor, and on this basis it appropriates the surplus labor of groups that have their own interests in maintaining their self-management functions as well as their authority, social position, and prestige.

This is the "pure" concept of state ownership as it appears in a socio-economic and political structure based on an all-embracing control of political authority in all spheres of the society, and on an established bureaucracy and a materially dependent and politically and ideologically subordinated and restricted working class. However, state ownership still characterizes the first phase of development in all countries which have taken the road of socialism as a possible, a necessary, and a progressive approach to socializing the means of production and as the basis for a more rational and planned type of economy. In a specific sense, state ownership is the elementary form for socializing the ownership of the means of production, provided that at the same time it is also the basis for further and more real socializing of socio-political relations—the preparatory process of destatization and the "withering away of the state" leading to the establishment of economic and political sovereignty of the working people, and the gradual acquisition of worker's social self-management. However, there is no reason for assuming that statism and state ownership are under all conditions a necessary phase in the transition from capitalism to socialism or for assuming that state ownership is inevitably the primary form of socialist transformation and of socializing ownership. Such an attitude is scientifically doubtful; for the forms of transition to a new society—and therefore the forms of ownership—are different and dependent on previous history, on the degree of economic and cultural development, and on other circumstances of time and country in which the social transformation occurs.

State ownership of the means of production in theory creates a monopoly of economic and political authority, and in theory and in practice makes possible the unification of economic and political authority under the control of a social group which represents the state, regardless of its representativeness and merit or any historical justification of its holding a position of power. When such a unified authority takes over an ideological monopoly as well, institutionalizing itself as a strong state, ready and free to apply force ("each state," Trotsky claimed before Stalin and contrary to Lenin, "is based on power"), then any needs for choice beyond the "independent authority" and any potential adoption of alternative roads to the "transition to socialism on the social level" are put off and vanish. In such a system of institutionalized and crystallized state ownership, "politics" not only remains independent but also becomes omnipotent; the planning and leading of society are all-inclusive and beyond popular control by the workers: true political processes are reduced to personal rule, struggle for power, and manipulation of the work and freedom of the people. It is today very difficult to defend the claim that nationalization—that is, the abolition of private ownership of the means of production—even in the form of the most developed and powerful state ownerships means also the abolition of classes and all their possible implications.

SOCIAL OWNERSHIP

Social ownership of the means of production is today the fundamental institution of the social and constitutional system of Yugoslavia. The Yugoslav constitution and Yugoslav political theory define it in the following terms:

(a) Social property does not confer ownership rights, rather it implies a form of removal from ownership of the means of production and of basic social relations; the Constitution of Yugoslavia expresses this explicitly and introduces a nonownership legal term: "social ownership of the means of production" (instead of *over* the means of production).

(b) It designates the *new* characteristic of the means of production as unowned, social: they are "the people's tools of labor" (Marx) and thus "returned to the producer" and inalienable in the socio-political sense, i.e., they cannot be the object of legal private or state appropriation, except if certain of them as objects are subjectivized for the satisfaction of personal and general needs.

(c) Social property expresses a new social relationship among workers who dispose and utilize the means of production and who are associated under the principle of self-management, both in the micro-society of their working organization and in the macro-society of democracy.

(d) Its socio-political base allows for relations of working and social solidarity, not of subordination (hired-labor) and implies the right of every worker to participate in self-management and in the new mutual working-relations which are guaranteed against the phenomenon of domination and exploitation of those who work by those who manage; every producer is also a manager, every manager is also a worker, which presupposes the transcendence of the contradictions between ruling and the executing functions (the antagonistic division of labor).

(e) It presupposes economic motivation: personal and collective interest in economic operations, in socializing the means for social production, projection, and planning of the general social development, in coordination, integration, and common decision-making on the part of the working people within the framework of self-managing economic, scientific, and political delegations.

(f) In functioning as the basis for resolving the dualisms between the individual and society, between ownership and non-ownership, and between command and execution, social property alters the concept of ownership rights by transforming it into a relation of responsibility toward one's self, toward the next worker, and toward the community. It is the highest possible measure of the equality and freedom of man in the given historical situation, engendering social control as a new function of guiding, developing, and guaranteeing responsibility as a new principle of personal and collective self-management of the individual, of the collective, and of the entire social community.

(g) Realizing itself in its content, social property at the same time negates and loses itself, transforming itself into a common possession of the means of labor that loses all institutional character even in personal ownership; an ownership that is reflected in the active appropriation of consumer goods within a medium constituting freedom of the labor and the personality of man, who now looks upon the social means and institutions as his own and upon the human environment as a vehicle for meeting human needs, for creative activity, and for enjoyment.

SOCIAL OWNERSHIP AND RIGHTS

Although it is neither private nor state nor any other kind of individualized ownership right, social ownership is not beyond objective law or even subjectivized law except patrimonial law. The relationship between social ownership and law is fundamentally and relatively new, distinct from public law on the one side and from private law on the other. It is primarily but not exclusively a category of constitutional law in socialism: in fact, every ownership is that, directly or indirectly. Bourgeois constitutions understand it as the basis of social relationships and formally guarantee (and later limit) it as private ownership rights. The first socialist constitutions introduced state ownership over the means of production as a global description for the new basis of socialist states. However, the Yugoslav Constitution of 1963 formulates social ownership as a new historical form of decapitalizing basic social relationships and restoring economic and social organization to the producers and to their "associates," and thereby constituting the basis for the common labor of people and for the liberation of labor as the necessary condition for and guarantee of the demonopolization of economic and political power and the creation of a new nonpossessive relationship: self-management. In addition, the Constitution of 1963 guarantees social ownership against parcelling into pieces of private ownership as well as against statist or collectivistic monopolism—also so-called cooperative ownership is defined as social. It attempts to work out certain new constitutional aspects and elements concerning social ownership. All of this is not always carried out clearly and completely nor decisively and substantivally, but it is carried out to the degree that it is now possible to provisionally determine the logical-legal concept of social ownership and to the degree that some conclusions can be drawn from this experience for further legal elaboration of these new social categories.

THE RIGHT TO SELF-MANAGEMENT

Social ownership, as has been pointed out, is a concept-process and consequently its structure is compound, pluralistic, and diffuse. It cannot be expressed, as once could private and later state ownership, as general or global law analogous to Roman private ownership—no matter what kind of nonownership concept is proposed: "the right to manage" or "the right of use" which is still persistently being used—even though only as a *terminus technicus* —in Yugoslav legislation, "the right of disposal," "quasi-ownership," "semi-ownership," etc. The concepts of Justinian's Code or Napoleon's Civil Code cannot be applied to it, and particularly not their abstract and monolithic concept of ownership rights as "absolute" ownership over a thing. Once ownership is shattered as a guise of power, an entire fan of rights germinates from social ownership; these however are not "real" but functional, not private but social and individual (communal and personal) rights implying authorizations; rights which terminologically and conceptually may be ancient but are nevertheless completely new.

Social property engenders also obligations, and in particular it entails a new form and meaning of responsibility. All of these rights, obligations, and responsibilities are *elements of the right to self-management.* They belong to the new legal branch of self-management which, on a new basis with a significantly new meaning, still maintains many norms of the earlier "civil law" and "working law," but transforms these branches radically. It creates new legal categories and legal relations—and is called to do that more often and better than has been done so far.

All of these rights are derivative and relative, and when they acquire the strength of subjective rights ("individual rights") then it is done within the framework of self-management and ownership as a *social service* up to and not beyond the point past which legal subjectivity would mean the destruction of the unity of society engendering monopolism, arbitrariness, and noneconomism.

Social ownership creates new rights: the right of working organizations to income, the right of working cooperatives to distribute that income according to work, the right of every man to his personal income, and the right to organize working relationships independently, are equally derived from social ownership. On the other hand, social planning, and the coordination of social activity in the general direction and organization of economic development, is a new right of social cooperatives. This "right-function" is not founded on political power even though some of the more important functions are still not separate from it. As a result these rights are derivative, relative and functional, and they are provided, protected, and limited by the Constitution, as are those of economic enterprises and other cells of associated work. The Constitution confirms that all of these rights are at the same time duties (commitments). Commitments of this sort are not only a proof that individual rights resulting from social ownership are not owned privately or innate but are rather conceived as social-and-one's-own at the same time. Some of these rights-commitments are expressed also through positive law: their execution is controlled and protected by the court; they are guaranteed in various forms of responsibility, from the moral-political and economic to the penal and criminal, and from the individual to the collective in the responsibility of the enterprise for economic violations and infringements.

PROBLEMS IN MANAGING SOCIAL PROPERTY; IN PARTICULAR THE SOCIAL PROPERTY OF MANKIND

1. There are some analogies between the concept of the "common good" or "general good of mankind," however diffuse, and the classical one of "general belongings" and "things out of circulation" *(res extra comercium)* in municipal law. This concept applies to the natural appropriation of specific elements of nature (air, etc.) or goods which are universally used (roads, parks, etc.). In principle all of these objects are beyond direct economic usurpation. This concept presupposes a negative form of freedom

from subjective (private or state) exploitation without the rights of modification and disposal. In concrete social and international conditions such a concept not only leaves open the question of the regime over the use and general management of these goods but also entails the possibility of great inequality in appropriation and distribution, so that the most developed and technically strongest nations may introduce virtual ownership and monopoly over such so-called general or common goods.

Social ownership refers to the means of production and to other means of work, as well as to certain parts of nature which are the conditions for work and production and thereby create new values. It negates all real and virtual ownership rights including the right of possession on the part of those who have the means to use particular goods for their own purposes. Social ownership presupposes the right of each individual and therefore of the common regime which comprises everyone; it excludes partial and private exploitation on the part of specific social groups and countries. It is the development and concretization of the old concept of the common good of mankind and thereby represents a concrete and efficient form of the internationalization of certain means which appertain to mankind as a whole and over which mankind as such—as an aggregate of equal people and countries—is the only entity to have the right of "social control" in the full sense of that concept.

With this in mind, the concept of social ownership is organically tied to the concept of management. Social property and its complementary regime of management represent a theoretical, political, and legal whole. In a functional sense, management implies specific actions with respect to the maintenance, conservation, and use of the objects of social property, i.e., an economic-technical and social usage which presupposes a corresponding regime of investment and distribution of products obtained by the economic-technical use of the means in social property. Politically, managing means not only administering, transacting, and conserving but also planning, developing, and distributing. All of this calls for a special socio-legal regime which needs specific definitions. Furthermore—and this is the political aspect—that regime must include machinery for management which guarantees integrity, social function, and social use of those objects which enter the regime as social and thereby international social property.

2. In the history of socio-political theory and particularly in contemporary socio-political and legal practice various forms of workers' and half-workers'-half-owner's management, i.e., self-management are known. In that respect, Yugoslavia has the widest experience, particularly after the introduction of workers' self-management two decades ago. Workers' self-management, which introduced not only workers' control but also the right of decision on economic and social problems and on relations of economic enterprises, developed gradually into a system which is called in this country "social self-management."

Social self-management has several meanings which are of interest for constructing a model of self-management concerned with objects and means which are or which should effectively become social property—the social property of mankind in its full sociological-political sense.

Fundamentally, social self-management means that not only in the economic but in every sphere of activity (all organizations based on common labor and therefore on the social ownership of the means of work), management, as an inalienable right, belongs to those who work in those organizations and institutions, to the working communities, the producers of both material and spiritual goods. The logical justification of the term social management is based also on the fact that in all fields of wide or public interest (education, national health, science, etc.) interested citizens participate in the management. These citizens are in principle delegated by the presidium of the municipal, republican, federal bodies, that is, the territorial-political communities which execute the functions of authority and management of wide or common interests in the territories which in the political sense act as territorial states.

Finally, social self-management implies a new mechanism for decision-making on common or general questions of management in fundamental aspects of work and social activity. This, in turn, transforms the traditional representative political structures and creates a new synthesis of individual and common interests, of autonomy and unity. It is on this basis that, in Yugoslavia for example, all representative bodies are defined not only as organs of power but also as organs of general social self-management. They are constituted by the political house and the house which represents the producers of the self-managers in various fields of social work: including economics, social politics, health, education and science. These modified representative bodies issue economic and other plans, programs, laws, and other general measures. These are acts of management, for they relate both to the direction and the unification of activities and management in basic organizations of work, and to the establishment of unique, or by necessity differentiated, regimes which determine the status of these organizations and the dimensions of their independent management as well as the system of responsibilities which result from the plans, acts, and other general measures.

The incompletions and peculiarities of the concept of social management of social property that are, among other things, consequences of the degree of material and cultural development and other qualities of a country still in the process of development and transformation into an industrialized society and its corresponding civilization cannot be denied. There is, however, no doubt that the theoretical definitions and the essence of social ownership as well as the historical experience in which ownership has to be assured by the establishment of new social relations, new forms of management—and hence new concepts which transcend the classical concepts of ownership, of man, and of decision-making—must be of interest to those engaged in building a model for the managing of international social property. Several conclusions can in fact be drawn.

(a) It is impossible to create a system for managing the social property of mankind by taking the path of analogies to existing forms of social management.

(b) The management of these forms of social property must be adapted to the character and form of these means of social property as well as to the existing social and political structures in the world, and at the same time every effort must be made to increasingly satisfy the interests and needs of all nations and hence to transform certain traditional concepts of international relations such as relationships of force and power.

(c) Past and present experience with the institution of social property and social self-management can be an inspiration and encouragement for the formulation of theoretical international models of management and hence enhance international awareness and contribute to the popularization, acceptance, and progressive realization of these concepts in the interest of the general development of the world, of peace, and of other goals of the international community as proclaimed in the Charter of the United Nations and in other international documents.

3. The international regulation of the seabed and the seacoasts can be successful and can correspond to the concept of social property if present and traditional concepts of international relations as relations of power, as compromises on the part of governments, and as semi-artificial and semi-pragmatic superstructures which express and protect the existing social and political status quo are transcended. This regulation must be the occasion for international regimes to approach their real purpose as instruments for uniting and satisfying mankind as a whole. It can introduce new ideas which correspond to the needs of mankind, that is, of people and nations seeking participation and rights in the utilization and the use of the acquisitions of science and technology, and especially of those objects and spheres which naturally belong not to those who have power but to those who live, work, and fight to live better on the basis of participation in the achievements of mankind and the sources of nature that belong only to mankind as a whole.

Starting from this hypothesis, certain assumptions can be made with regard to possible models. Direct management, that is, utilization, use and conservation, plus all of the economic and legal consequences this would entail, cannot be entrusted to a single organizational mechanism which would be a monster international enterprise. The technology of work and other peculiarities of the seabed, its geographical position, and the problems of efficient management will demand in principle regional and similar enterprises for direct management. However, it is an inevitable consequence of the concept of social property that these enterprises cannot be national, by proxy, or mixed, meaning an organization of "interested" or territorially national states. They can only be enterprises of social property and hence social enterprises of mankind.

Certainly, the conditions are not ripe at this time for spelling out the mode of selection of workers, technicians, and managers for these enterprises; nor for proposing complete models determining their positions, or the rights and techniques of management. But international social ownership irreversibly postulates an international regime to select cadres and to subject them to international control. It postulates in fact a *new type of man* who serves the needs of mankind with his own work regardless of his national origin or national citizenship. The concept of this new man transcends that of the existing international employee as an agent of international organizations. The new man is an "employee of mankind," working for all people in the world. These producers and managers furthermore must have a different role in decision-making, in the distribution of labor, in income, and in other social rights than they have in the majority of national or international enterprises. A dimension of social self-management is inseparable from any concept of social property and hence of the social property of mankind.

4. Social self-management is not a one-dimensional principle. It includes: (a) combined management on various levels of organizations of a particular activity; (b) various functions and degrees of decision-making. Developed social self-management is a *system* which guarantees the autonomy of the individual and the basic unity of action, the harmonizing of functions and mutual relations, and the responsibility of the subjects and their "delegations." This is a system that is built from the bottom up, from the people (in its contemporary complete and multi-dimensional being), a system that is integrated in order to prevent

"autonomization" (monopoly and group selfishness). All of this is of special significance for constitutionalizing this ocean regime system of managing the social property of mankind.

On the international plane general management might belong to the Maritime Assembly. The Maritime Assembly might have three houses: political, professional, and scientific. These would be equal and might be composed of a hundred members each. The first house should represent the state on a basis of regional representation, and might be elected by the General Assembly of the United Nations on a basis of candidates nominated by the national assemblies. The second house should consist of delegates from professional workers' and consumers' international and national organizations. The third house should be made up of delegates from scientific and technical organizations and eminent individual scholars. Initially, the members of the latter two houses might be elected by the General Assembly of the United Nations on recommendations of its competent commissions and agencies. Working people engaged in maritime organizations for the exploration and exploitation of ocean resources should be entitled to send five delegates to each of these two houses.

The Assembly should be the highest organization of management and decision-making in questions of planning, utilization, investments, distribution of means and products, and other important political, scientific, and functional questions. It should also be the arbiter in the last instance. It would elect the Executive Committee. The President of the Committee and four secretaries, selected from the ranks of competent experts, would form the Maritime General Secretariat which should execute the current managerial affairs and prepare meetings and the work of the Committee. The Committee should do this for the Assembly. The Committee and the General Secretariat would be responsible to the Assembly, but the Secretariat would work on the advice of the Committee.

This management machinery should be autonomous and independent. A special fund for financing its work should be provided by the maritime organizations in agreement with the statute confirmed by the Maritime Assembly. Working collectives and any five states or particular international organizations could seek the recall of any member of the Assembly and other organs which inflict harm on social property, on management, or on other basic interests of mankind confirmed in the Constitution of Maritime Economy.

It is not impossible that intermediary machinery will be established on the level of specific oceans, seas, regions, continents, etc. This machinery would have certain independent functions of management and could even carry out some of the work of the Maritime Assembly in its domain. It would be charged with the execution of certain decisions and jobs by the general international management machinery.

In planning the composition, the methods of work, and the decision-making procedures of these organizations of self-management, it is essential that every precaution must be taken in order to prevent infiltration by imperialistic tendencies as well as phenomena of bureaucratism. To proclaim social property, and even to establish the most rational and unique system of management, does not automatically assure the realization of the best interests of mankind, nor does it automatically guarantee immunity from the manipulation and exploitation of the stronger or of the managerial technocrats who can establish themselves as a new power.

All this requires constant public and scientific control by international public media and, above all, changes in the existing international constellation in which power and force determine relationships among the big powers and between them, jointly or individually, and the others.

Social property and social management of the common wealth of all people and all nations demand a transformation of relationships of inequality and domination. These changes are inevitable, for there can be no durable peace or lasting solution of any problem on earth without the participation of all and without respect for the interests of all. In addition, a large section of mankind is today socialized because it is existentially interdependent. Without the international socialization of sources of energy, this segment of mankind cannot live in peace and develop its wealth. In essence, this means that not even mankind as a whole can continue any longer, without risking deep conflicts and their consequences, unless it *shares* that which is by nature and by rational function common-social property. This is particularly true with respect to the sea and the wealth in its depths and its bed. Establishing solidarity, cooperation and reciprocity is one of the conditions for the prevention of new divisions leading to new conflicts and to great catastrophies.

The social property of mankind and its management is not only an essential technical-economic question; it is all also a moral-political problem of life importance for the world and for each one of us.

FRANCIS T. CHRISTY, JR.

12 Fisheries: Common Property, Open Access, and the Common Heritage

The term *common property* is often used in references to the living resources of the high seas. This use of the term is misleading because it incorporates two quite different concepts. One, that of *res omnium communis,* implies that the resource is jointly owned or held in some indistinct fashion by all members of a community. The other concept, that of *open access,* means that the resource cannot be used exclusively by one or a few members of a community, but is open to free use by all. The first concept concerns ownership and relates primarily to issues about the distribution of wealth. The second concerns use and relates primarily to the production of wealth. Thus far, international fishery arrangements have failed to make clear distinctions between these two concepts of common property. If the confusion continues and the distinctions are not made clear and operable, the sea's wealth in fisheries will continue in the duality of appropriation by a few states and dissipation by many.

The trends toward both appropriation and dissipation of wealth are already in evidence. Appropriation is manifested not only by an increasing number of states in the extending of their limits of fishery jurisdiction, but also by the increasing number of attempts to divide fishery yields among a few participants. The dissipation of wealth is seen both in the growing number of depleted fish stocks and in the congestion of vessels on the fishing grounds.

FRANCIS T. CHRISTY, JR., is with Resources for the Future Inc., Washington, D.C.

The severity of the problems associated with these trends and the severity of the difficulties of overcoming them are closely related to the two different concepts of common property. If the first concept, *res omnium communis,* is not strongly and widely held, then the difficulties of appropriation are not particularly great. If the second concept, open access, is not perceived as being damaging to the total range of the interests of participating states, then the costs of wealth dissipation are not particularly great. Under these conditions, common property might continue to be an acceptable guide for the management of international fisheries. But before making that choice, it is critical that states examine as closely as possible the consequences to their interests of operating under these two concepts.

The examination cannot long be delayed. The re-opening in the near future of the Geneva Convention on the Continental Shelf is inevitable. And although attempts to restrict discussions to the resources of the seabed are likely, these will be unsuccessful because of the inextricable relationships among the different uses of the seas.[1]

It is also questionable whether or not the discussions *should be* restricted to the seabed. As Stuart Scheingold has pointed out, the success of an ocean regime will in part be dependent upon the ability of the regime to comprehend converging interests, and this can be done best by allowing for a wide variety of trade-offs, not just those related directly to the bed of the sea.[2] Enlarging the scope of discussion permits more explicit recognition of the diverse items that will be integral to the bargaining process.

In this context, it is important that States begin to determine as precisely as possible their interests in fisheries as well as in other uses and resources of the sea. But fisheries are also important *per se.* While there are few, if any, state economies that are significantly dependent upon fisheries as a source of income, the catching of fish is the only enterprise (other than war) in which states are in direct confrontation with each other over the same resource. The conflicts emerging from this confrontation are growing in severity. Thus, apart from discussions of an ocean regime, a re-examination of the concepts of common property may be inevitable in the near future.

DEFINING RES OMNIUM COMMUNIS: A FUNCTIONAL APPROACH

Attempts to elaborate the *res omnium communis* concept of common property generally lead one quickly into a morass of ill-defined legal and social ownership concepts and terms. Endless etymological arguments can be—and unfortunately, are—waged over definitions of private and public property, social property, public goods, title, ownership, community interest, and the common heritage of mankind. But while these arguments may delight and employ legal scholars, they are of little immediate relevance to the resolution of the issues of the sea. Much more important is the practical determination of the way in which states perceive their interest in the uses and resources of the marine environment. And this is not a matter of defining ownership or common heritage but of determining who gets what, or more precisely, the extent to which states feel they have a share in the wealth of the seas. Decisions will not be made on the basis of private, public, or social property, but on expectations of wealth.

PERCEPTIONS OF WEALTH

Wealth, in this context, should be broadly defined to include all items *(res)* of value to states; economic revenues, power, security, prestige, employment opportunities, food, etc.[3] Fisheries are a source of several of these items, the most important of which is the purely economic one, a value that can be measured in terms of its contribution to the state's economy. But other values, more difficult to measure, are often sought as well. These include the provision of opportunities for underemployed labor; self-sufficiency in the production of food; production of food for the good of mankind; and as sometimes stated even preservation of maritime skills for use in the case of war.

The value a state gives to these items will affect its concept of the commonness of common property. The greater the perceived value, the more likely it is that a state will want to share in the distribution of wealth: the lesser the perceived value, the lesser the interest.

Schaefer, who has studied the estimates of potential yield made by a number of scientists, has concluded that "the attainment of fisheries productivity of two hundred million tons per year is realistic, and doubtless conservative." [4] These figures are all compared to the 1968 global catch of sixty-four million metric tons (of which 6.7 million comes from freshwater sources).

As discussed below, these estimates are of questionable relevance to the consideration of the sea's wealth since they are derived largely from physical concepts modified only slightly by intuitions of economic costs and revenues. But more importantly, they may actually lead to misperceptions of wealth.

Phrased in different terms, the sea's wealth in fisheries is not a function of physical quantities but of perceived economic scarcity of the resource *in situ.* If the resources are perceived as being so abundant that all can win and none will lose in their enjoyment, then there is no value in exclusive rights to the resource itself. But if resources are perceived as being scarce, so that use by one fisherman diminishes use by others, then there is value in having exclusive rights. It is on the basis of this value that the wealth should be estimated for this is the source of economic rent and also the source of conflict among states.

NON-INCLUSIVE PATTERNS OF WEALTH DISTRIBUTION

It is assumed here that most presently-utilized fishery resources of the ocean are scarce economic resources and that they constitute items of wealth. Questions can then be raised as to the degree to which this wealth is common under the present regime and as to the degree to which states may expect to share in the distribution of this wealth in the future. The answers to these questions will help to characterize the common heritage concept of common property.

Under the present regime beyond the limits of national jurisdiction there are two elements that legally determine the distribution of wealth in fisheries. First, the principle of freedom of the seas provides that all states have free and equal access to all fisheries in international waters. Second, the only way in which wealth can be obtained is by the exercise of this right of access—

only those who actually fish share in the distribution of fisheries. There are, however, significant exceptions to these elements which have an important effect on the patterns of distribution. Moreover, trends apparent in international arrangements and unilateral actions are exacerbating this alteration toward nonexclusive patterns of distribution and toward the diminution of the degree of commonness of common property. This can be observed in several different situations: open access, national quotas, abstention, limits of exclusive rights, preferential rights, and economic efficiency.

Open access

The condition of open access itself has a distributive effect. It might be assumed that the distribution is proportionate to fishing effort; so that, for example, a state with ten percent of the vessels would get ten percent of the catch. But this is only partly true, because there are large differences in the kinds of effort and their management, and these have significant effects on the amount of catch.

Three examples of disproportionate gains in wealth can be described. First, some states have no control over the number of vessels participating in a fishery. As noted below, this tends to induce excessive amounts of effort to enter the fishery; so much so that the economic wealth in the fishery tends to become dissipated. While the absence of controls on effort may permit the employment of larger amounts of labor, it does so at the sacrifice of returns to labor and to the economy of the state. Thus, for similar amounts of effort, these states receive lesser shares of economic wealth than do those that have controls over the number of vessels.

A second example relates to differences in the kinds of effort. Certain states have large distant-water fleets. These are highly organized and centrally directed. But more important, they have a great deal of mobility and adaptability. When they fish off the shores of foreign states, they are frequently able to do so more effectively than the smaller local coastal vessels. If they deplete the local stock, their mobility and adaptability permits them to move on to other stocks off different shores, but the small local coastal vessels must remain to fish the depleted stock, and their access to wealth is consequently diminished.

A third example results from certain kinds of conservation regulations. One of these is the establishment of a limit or a quota on the total amount of fish that can be taken from a stock during a season. This kind of regulation stimulates fishermen to build larger and faster vessels so that they will be able to get greater shares of the fish stock for themselves before the total quota is reached and the season closes. For example, the Inter-American Tropical Tuna Commission instituted a total quota on catch of yellowfin tuna in 1967. When the quota was established, the capacity (total tonnage) of the United States tuna fleet was thirty-seven thousand tons. This amount increased very rapidly in subsequent years and by the end of 1970 the capacity is expected to reach sixty-five thousand tons. This seventy-five percent increase in four years is made up almost entirely of larger and faster vessels. This kind of vessel will get the greatest share of the fixed amount of catch, while the older and smaller vessels will have a much smaller proportion than they did previously. Since the United States is better able to invest in larger and faster vessels than are the other signatories to the treaty, the distributive effect operates against the other parties, as well as against United States owners of smaller vessels.

These examples indicate that even under the conditions of open access and nonexclusive arrangements, the distribution of wealth is not proportionate to the amount of effort invested, but is affected by differences in management, by differences in kinds of vessels, and by variation in conservation regulation. The distributive effect is not necessarily in favor of the developed states. In the first example, the negative effects are quite significant for the North American and Western European countries. In the second example, the advantages lie not only with the Soviet and Japanese fleets but also with the fleets of some of the developing states, such as South Korea, that have been able to mount distant-water enterprises. In general, the states able to employ highly mobile and tightly organized fleets are receiving greater relative shares of the sea's wealth in international fisheries, while those not so able are receiving much smaller—often negligible—net returns.

National quotas

A second, and more important, form of distribution occurs when open access is subjected to certain kinds of constraints. One of these occurs in international agreements that divide the annual yield among a few parties. There are several such agreements.[5] Two of these are described briefly in order to indicate their dependence upon the exclusion of nonparties.

The first international conservation agreement was signed in 1911 and is still in effect. Under this agreement among Japan, the Soviet Union, the United States, and Canada, the catching of fur seals on the high seas is prohibited. Instead, all harvesting is done on the breeding islands by the states that have jurisdiction over the islands—the Soviets and the United States. Canada and Japan, which have given up their freedom to take seals on the high seas, receive in return fifteen percent of the pelts produced. Economically, this is the most efficient of all international agreements because it permits the harvesting to take place at the minimum cost.

Another arrangement—and one far less successful—is that of the whaling nations participating in the Antarctic. Several years ago, five states came to an agreement among themselves on the number of blue whale units (a unit of measure covering the various species according to size) that each would be permitted to take during each season. Unfortunately, the only agreements that they could reach were based upon totals that were beyond the capacity of the stocks to sustain. Thus depletion has become more and more severe. Nevertheless, the shares attached to the fleets of the various states quickly acquired additional value, so much so that the Japanese were willing to buy the fleets of the British and the Dutch not for the vessels but for the shares of the total quota that were attached to them. In this case, the de facto right to take whales has been sufficiently strong to permit a market to develop for the shares.

But as attention is brought to bear on efficiency questions and on such devices as national quotas, it is inevitable that related questions of wealth distribution will also arise with increasing force. As Oda has noted with respect to Japan's interests in the North Pacific, "we discover that the traditional formula of free competition of fishing on the high seas has been replaced by different formulas for allocating these fishery resources." [6]

The success of a national quota scheme is dependent upon the ability of the parties to exclude new entrants. If a state not party to the agreement begins to fish the managed stock—as it has every right to do under the principle of the freedom of the seas—then either more fish must be found or the shares of the participants must be lowered. The former is unlikely because the fishery would not be under this form of management if it were not fully utilized and in danger of depletion. Thus, any increase in total permissible catch would defeat the purpose of the agreement. The alternative of lowering the participants' shares in order to accommodate the new entrant would be extremely difficult. It is more likely that the signatories would bring sanctions to bear against the new entrant with the intention of forcing him out of the fishery.

The severity of the problem is self-regenerating. If national quotas are adopted and the parties rationalize their fish-catching industries, the values of the shares will become significant. These values will increase in response to growing demand, a factor that can be anticipated for most of the fisheries under management. And as the values increase, the nonparties will have growing incentives to enter and the signatories will have greater desire to exclude them.

Several suggestions have been made for dealing with the problems of new entrants.[7] One of these is to leave aside a certain share in the original distribution for new entrants. This, however, simply postpones the eventual necessity of dealing with the problem. If the size of the reserved share is small, it might be taken up by one or by two new entrants and the third entrant would still have to be excluded.

The trend toward the adoption of national quotas parallels the trends toward the extension of exclusive fishing limits and the use of discriminatory conservation regulations. As fish stocks become fully utilized—and possibly depleted—it is clear that new entrants will reduce the fish-catching potential of present participants. Under these conditions, it is natural for the "insiders" to seek means of excluding the "outsiders" and of acquiring as great a share of the annual catch for themselves as they can. If a state cannot invoke other techniques that are more favorable to its interests, and if it becomes clear that it can no longer increase its own share of the catch, then it is likely to advocate the imposition of national quotas.

For example, in the North Atlantic the catches of cod and haddock have reached the maximum and may actually be producing lower yields than they could under proper management. Though total catch cannot be increased, some states may expect to increase their proportion of the total by employing more vessels. Such states will not be willing to agree to national quotas on the basis of past records of catch and will wish to postpone division of the quota until they feel they can no longer increase their catch. As they reach this position, they will come to accept the national quota device.

The basis for the division of the sea's wealth through national quotas is the concept of historic rights. It is assumed that a state acquires a share of the resource by virtue of having exercised the right to fish. This assumption can be considered a moderate form of the doctrine of occupation under which states acquired jurisdiction over land areas in the past. Or, if the technique becomes more prevalent, then historic rights to high seas fisheries may assume the mantle of customary law.

But expectations of states are more important than legal descriptions of claims. And at this point in time it is not clear whether the majority of states expect that they can acquire wealth in high seas fisheries only through a history of use or whether they expect that that wealth is available through other means reflecting the more inclusive concept of common heritage.

Abstention

The doctrine of abstention is another, although far more restricted, technique that has been used to affect the distribution of the sea's wealth in fisheries. Under this doctrine, parties to the agreement abstain from entering a particular fishery where that fishery is fully utilized by other parties and is subject to conservation efforts and scientific study.

The only present example of the abstention doctrine is the International Convention for the High Seas Fisheries of the North Pacific Ocean, signed in 1952 by Japan, Canada, and the United States. The effect of the agreement has been to prohibit Japanese fishing for salmon in the eastern half of the North Pacific, east of the 175° W meridian. Oda comments that "the abstention formula ensures a maximum share to one party, while giving nothing in return to others. The writer does not hesitate to count the 1952 Convention as one of the most epoch-making treaties in the history of international law, in that it broke the previous conceptions of the exploration of marine resources, namely, that all nations concerned should compete with each other on an equal basis to acquire as many resources as possible within limitations (applying equally to all) aimed at conserving those resources." [8]

Through the imposition of this doctrine, Canada and the United States gained exclusive enjoyment of the salmon stocks of the eastern North Pacific; an enjoyment, however, dependent upon the willingness of nonparties to refrain from entering. There are currently considerable apprehensions about possible encroachments upon the stocks by the vessels of South Korea.

In a report of this problem, one quite interesting observation emerged. It was suggested that "the Koreans feel they are in a special position inasmuch as their fishing fleets are partly set up by the United Nations." [9]

More recently, the doctrine has received additional attention for the salmon of the North Atlantic. This has occurred because of the recent discovery of aggregations of salmon off the east coast of Greenland. These aggregations include salmon originating in streams on both side of the Atlantic, and the taking of them by Danes and others has apparently been sufficiently successful to reduce the numbers returning to the spawning streams. Sports fishermen, with heavy investments in salmon beats (stream rights), have protested the taking of salmon on the high seas and away from the spawning streams. In 1969, both the Northeast and Northwest Atlantic fishery commissions voted in favor of abstention rules that would prohibit the taking of salmon outside territorial waters. Whether or not the resolutions can be brought into effect remains to be seen.

The abstention doctrine employs a principle other than and different from historic rights in the struggle over the sea's wealth. This other principle is particularly applicable to stocks that are susceptible to protection or enhancement by actual investments in management devices, such as salmon. For example, the construction of fish ladders on dams, the development of hatcheries, and the employment of pollution controls all create costs that are borne by the state in whose streams the salmon spawn. Unless these states can be assured of a fair return on their investments, they will be unwilling to undertake them. They assume—with some justification—that they have some right in the salmon even when they are in the high seas.

On the other hand, the enforcement of such rights through the abstention principle permits the appropriation of the wealth by the spawning states and excludes those not so favored. The excluded states could point out that salmon in the high seas are grazing on the common property of the world community and using materials otherwise utilizable by other species. In this sense, they could claim that they also have a right to share in the resource. But such a concept has not yet been advanced and the abstention doctrine remains as an important element in the noninclusive distribution of wealth.

Limits of exclusive rights

The extension of exclusive fishing limits by coastal states into waters farther from their shores has become increasingly prevalent in the past two decades. A limit of twelve miles is now common, but it is by no means fixed. Claims of much greater distance have been made and are enforced, and more such claims can be expected as the fisheries adjacent to coastal states become more valuable and attract more fishermen from distant shores. These extensions clearly reduce the common heritage, however that is defined, and they distribute the sea's wealth in favor of those states with long coasts and valuable stocks.

Preferential rights

As a means of reducing and avoiding proliferation of such far-reaching claims, the United States is attempting to develop a system based on preferential rights of coastal states. This system would recognize the special interests of coastal states in resources adjacent to but outside its twelve mile limit. The preferential treatment would be accorded only to those stocks that require management and in which the coastal state has demonstrated an interest. The system would not, therefore, exclude foreign fishermen from taking fish that the coastal state has not fished.

In developing this system, the United States has reached a number of *ad hoc* short-term bilateral agreements with Soviet Russia, Japan, Mexico, and Poland.[10] The arrangements are quite complicated—permitting the foreigners to engage in certain activities within the U.S. twelve mile zone in exchange for their restriction of or actual abstention from fishing for certain species at certain times in areas outside the twelve mile zone. Some of the zones are more than a hundred miles from the coast of the United States.

While this system does not claim exclusive rights to areas of the high seas, it does assert the principle of preferential rights. It attempts to provide or to ensure the coastal state a share in the sea's wealth that it might not be able to acquire under open access. It is likely that new entrants would have the same difficulty participating in a fishery subject to preferential rights that they would have in one subject to national quotas.

Economic efficiency

I have made several references to the open-access concept of common property and its deleterious effects on production of economic wealth, and, while there are no international fishery arrangements—with the possible exception of the North Pacific Fur Seal Convention—that take these economic effects into consideration, the adoption of such arrangements would have a significant effect on patterns of distribution.

Free and open access means there are no controls on the amount of capital and labor that can be employed. This condition exists for a variety of resources both within and between national economies. In certain states, the United States for example, there is free and open access to common bodies of water, to public recreational areas, to most sport fisheries and wildlife, to air, and to many other resources. Formerly in the United States common pools of oil could be drained by anyone with rights to the surface above. In other states some of these resources are open and others are closed, but this condition is not a function of the political system governing the state, for it can exist in both market and centrally-planned economies. It is not a matter of differences in the concepts of property but of differences in the ways societies choose to manage their resources.

The condition of free and open access also exists for resources lying outside state jurisdiction. These include, or have in the past included, such resources as outer space, the radio spectrum, airspace, the atmospheric envelope, and resources and uses of the high seas. Generally, the condition exists in these areas because of the absence of jurisdiction and the difficulty of acquiring exclusive access.

If the resources are so abundant that use by one entity does not affect use by others, then the condition is a matter of little consequence. But when interference occurs, the maintenance of the condition can lead to considerable difficulties and waste. For the radio spectrum, for example, it has long been accepted that access must be controlled, for two units using the same frequency in the same area would lead to obvious inefficiencies and costs. Similarly, controls have been established on access to common pools of oil, to commercial airspace, to the uses of common bodies of water for the disposal of wastes, and to many other common-property resources.

But the necessity for controls in the above examples has not been accepted for the international use of marine fisheries. The condition of free and open access is maintained and the consequences have become both apparent and significant. The inability to control access has led to physical waste in the form of depletion of stocks, and to gross economic waste in the form of excessive application of capital and labor.

In physical terms, the waste occurs because no individual fisherman or fishing state can afford to restrict catch in the interest of future returns, for anything left in the sea for tomorrow will be taken by others today. As long as there are no controls and the demand for the product continues to rise, depletion is inevitable. It has already occurred for a large number of stocks and is increasingly prevalent.

But at some point the *increase* in the total costs brought about by the number of fisherman operating is greater than the *increase* in gross revenues they provide. This is almost simultaneous with the point at which there is the greatest difference between total costs and total revenues and the greatest profit to the industry. If the resource were under the management of a single agency that could control the number of vessels and fishermen, the agency would stop the addition of capital and labor at this point, prior—if possible—to the significant total-cost increase.

There have been only a few studies measuring the amount of waste taking place in fisheries. But each study reaches the same conclusions. In the Pacific salmon fishery of the United States and Canada, it has been estimated that ths same annual catch (and total revenue) could be taken with about fifty million dollars less of capital and labor than are currently employed each year.[11] An estimate of the Georges Bank haddock fishery made several years ago (prior to the utilization of that stock by the Soviets) stated that "the point of maximum profit would be at a level fifty percent or less of the recent average (amount of effort)."[12]

This consequence of economic waste in fisheries can be avoided only by providing some kind of controls over access that would exclude the superfluous vessels and fishermen. Controls can be established by taxes, license fees, auction or lease of rights, or other similar means. Fishermen who are willing to pay the fees or who obtain the licenses will be no worse off (and probably much better) than they were before. They would benefit by prevention of congestion and wasteful competition. The yield of the resource could be maintained at the maximum level. Society would benefit by receiving a share, or the full amount, of the surplus profit.

However, the system would require acceptance by the world community of a principle contrary to the freedom of the seas—that is, that access would no longer be free and open. There would in addition be a considerable number of transaction costs in the development and establishment of the immensely complicated arrangements required. Nevertheless, some system of exclusive access would permit a major increase in the production of economic wealth and one that could be shared inclusively.

Many of these devices (excluding the last), have been adopted in the name of conservation. While there is some question of the merits of conservation *per se,* the acceptance of it as an objective requires that all participants in a fishery follow the same rules and regulations. This tends to establish a precarious balance among the parties, each taking roughly the same proportion of the yield that it did in the past.

The entry of new parties usually disrupts the balance and there are no countervailing forces that will naturally restore equilibrium. The tendency, therefore, is to resist all new entrants and preserve the fishery for those who have established historic or preferential rights. If fishing states become more successful in resisting admission, then the value of the option to fish is diminished, and outsiders are excluded from sharing in the sea's wealth in fisheries.

The severity of exclusion depends in part on how the outsiders presently value their option to fish. The smaller, landlocked states may not view loss of access with great misgivings. However, other states may. For example, when Iceland, Norway, and Denmark extended their limits or those of their possessions, the United Kingdom experienced considerable loss. The 1961 loss of catch to British fleets was estimated to "range between twenty percent and thirty-five percent of the total annual catch from distant waters in recent years (ten percent to eighteen percent of total landings of white fish from all waters), plus nearly fifty percent of the middle water catch (roughly three percent of total white fish landings.)"[13]

But the United Kingdom, while an "outsider" in the above situation, is an "insider" in other situations. And this indicates the ambivalent position of many fishing states. As another example, United States fishermen who take tuna in the eastern tropical Pacific object to the two hundred-mile claims of Chile, Ecuador, and Peru, while the fishermen of the northeastern and northwestern United States would like to have a two hundred-mile limit at home in order to exclude "foreigners." If historic rights become more widely adopted as a basis for dividing the sea's wealth, fishermen from the United States might have gains in a few fisheries but would lose access to many more.

While for many states the net effect of current patterns of distribution is not clear, there are certain effects that can be generalized. The states with large distant-water fleets are acquiring significant shares of the sea's wealth because of their mobility and because their participation in many fisheries gives them historic rights. The states with long coastlines and valuable near-shore fisheries are also acquiring significant shares of the sea's wealth. They are doing this both by the extension of exclusive limits and by the claims of preferential rights. In all cases, however, states that do not control the number of their fishermen and vessels tend to dissipate the economic wealth of the shares of the resource that they acquire.

In short, the trends in catch appear to be leading to only a slightly more inclusive access to wealth and to only moderate increases in the shares of a few of the developing states. If the FAO projections of limited growth materialize, then even these slight changes toward more inclusive access will diminish.

Thus, under the present patterns and trends, the future opportunities for more inclusive access to the sea's wealth in fisheries are limited both by the conditions of demand and supply and by the prevalence of noninclusive arrangements and actions. Furthermore, as noted above, access must be limited if there is to be effective production of economic wealth.

These points should be emphasized. The trends in use balanced against the conditions of supply provide only limited opportunities for more inclusive access. The patterns of distribution and international arrangements serve to impede more inclusive access. And the possible adoption of techniques to prevent the dissipation of wealth will restrict more inclusive access. Therefore, *the distribution of the sea's wealth in fisheries will be severely limited and the common heritage of mankind will be significantly diminished so long as shares in the sea's wealth can be acquired only through the exercise of the right to fish.*

As this becomes more apparent, states may seek new techniques for participating more inclusively in the distribution of wealth. They may seek more precise definitions of common heritage which can serve as useful guides for the establishment of new international institutions and arrangements. The intensity and dimensions of the search for new meanings of common heritage will depend upon a number of factors: the number of states that feel that the exclusive patterns are more damaging than beneficial to their interests; the nonfishery trade-offs that may be important; the perceptions of wealth and of the different items that make up the wealth; and upon expectations of success in achieving demands.

This paper cannot discuss all of these factors, but it can suggest different ways in which more inclusive distribution can be achieved. An examination of these suggestions may help individual states define more precisely their understanding of the common-heritage concept of common property.

One system for sharing more inclusively would be to guarantee free and open access for all states to as large an area of international waters as possible. This would necessitate explicit recognition and reversal of the present trends toward exclusive arrangements and claims.

Another step might be the relinquishment of unilateral claims of extensive jurisdiction, so that the widest possible area would be open to all states. Third, more inclusive access might also be ensured by giving up the principle of preferential rights of coastal states. Fourth, limits on catches of major fishing states might also be imposed, or it might be decided that no state can take more than a certain proportion of the yield of any particular stock. Another example would be to actually subsidize fishing industries in nonfishing states in order to ensure their sharing of access.

Such steps as those above would help to preserve the principle of the freedom of the seas and to maintain as fully inclusive access to fisheries as possible. And, if wealth can only be acquired through access, these steps would help achieve more widespread distribution. But there would be a number of difficulties associated with adoption of such steps.

The political difficulties are manifest and need not be elaborated. Less obvious is the disastrous effect such a system would have on the efficient production of economic wealth. But even if the political difficulties could be overcome and the economic wastes were accepted, the system would still provide only a limited definition of common heritage. The distribution of wealth would be among only those states exercising their option to fish. The nonfishing states would be excluded, even though they would maintain their options. The sharing of access, therefore, might not be a satisfactory definition of common heritage.

A more direct system for sharing the sea's wealth in fisheries would be through the allocation of quotas to all states of the world. The maximum sustainable yield of each stock of fish fully utilized (cod and haddock in the North Atlantic, whales of the Antarctic, salmon of all oceans, etc.) might be divided up among the 120 some nation-states. Each state, therefore, no matter how small, landlocked, or uninterested in fishing, would have a certain quantity of fish that it could take each year. In essence, this would constitute a system of fully inclusive national quotas—and would have some of the same advantages and defects.

The major difficulty, of course, would be that of determining an acceptable distribution of shares. There would be a number of other difficulties attached to a system that divides up resources. The system would be cumbersome in view of the large number of "owners." The division of individual stocks into shares would be extremely difficult, particularly in those cases where the stocks spend part of their time within the territorial waters of some states, or when the stocks are ecologically interrelated with other desirable species. There may also be misallocation of effort among different grounds of the same stock or during different periods within the same season.

Nevertheless, conceptually at least, the direct sharing of yields from fishery stocks provides one definition of common heritage that is fully inclusive.

It is possible that a system could be developed that would distribute the commodities produced from the living resources of the sea. Shares would be in fresh, frozen, or canned fish, in protein concentrates, etc. Indeed, it is sometimes stated that states suffering from protein deficiencies should receive foodstuffs directly from those that produce them. Such a system might require each fishing vessel or each processing plant to give a share of its produce to some international distributing agency, which would then distribute the materials to the various states.

This system might differ from the previous one in that access to fisheries could still be free and open. Current kinds of conservation measures, including national quotas, might still be practiced. But a form of wealth would be extracted in the form of a tax consisting of commodities rather than money.

There would be considerable difficulties with such a system. Among others, would be the problems of determining the amount of commodities to be taken, the form of commodities, the allocation among states, etc. For example, there are wide variations among states in taste preferences and patterns of demand. Commodities may be rejected by some and welcomed by others on bases of fish size and species, processed form, container, etc. The task of allocation would be almost impossible.

As noted elsewhere, the condition of open access leads to the employment of superfluous amounts of vessels and fishermen. They are attracted into the fishery because the economic rent it produces can be shared by all participants. But as more fishermen enter the fishery, each one's share of the rent becomes smaller until it is reduced to zero and the total economic rent is dissipated. No one gets it, neither society nor the fishermen. The amount of rent may be quite large and, if it were not dissipated, it could become a source of wealth shared inclusively by the world community.

In order to prevent dissipation of rent, the condition of free and open access has to be removed and some form of exclusive rights has to be granted to the fishermen or to an agency managing the fishery. There are various means for achieving such rights and for reducing excess access, including taxes, auction or lease of rights, license fees, etc. Access could be kept open but not free. That is, anyone paying the fee would be permitted to fish.

The employment of such a system would require the establishment of an agency with considerable authority. It might be an agency restricted to certain stocks, such as Antarctic whales or Pacific salmon. It might be regional in nature, such as one for North Atlantic groundfish. Or it might be global, covering all fisheries of the oceans. But whatever its scope, the agency would have to have the power to extract the economic rent from the fishery. This would not reduce, necessarily, the amount of fish caught. Nor would it reduce revenues to the fishermen that remain in the industry. It would, however, prevent superfluous fishermen from entering the industry and it would reduce opportunities for employment. The rent extracted would reflect the value of the resource to the users. It would be similar to a grazing fee or to the bonus payments that oil companies pay to the federal government for rights to produce oil on the continental shelf of the United States.

This system, like the others, is not without its difficulties. There would be problems of determining the appropriate tax or fee, of determining the appropriate amount of effort and the kind of efforts, of allowing for an orderly rate of technological innovation, of allocating effort amoung grounds, between species, and within seasons, and several more.

An alternative distribution scheme would be to allocate shares of the rents directly to states on the basis of certain criteria such as need, population, area, etc. It would be quite difficult to determine and get universal acceptance of the criteria. It is also likely, for many years at least, that the shares would be minuscule. Nevertheless, a direct allocation of rents may be more acceptable than the establishment of a development fund.

The choice between the two would depend, in part, upon how states view their interests in the sea's wealth. Sharing in access to a development fund might be acceptable to developing states if they perceive the total wealth to be small and if they do not have strong expectations of participation. However, it may be that states will consider themselves as part owners of the sea's wealth and that they have a *right* to a certain share rather than simply the *privilege* of receiving benefits. In this case, they may prefer a direct allocation of rents and a distribution scheme that would be more inclusive and equitable than might be achieved through a development fund.

It is possible to conceive of a number of other ways in which more inclusive distribution of the sea's wealth might be effected. One of these would characterize the wealth not so much in terms of revenues or products but in terms of participation in decisions and management. Most international fishery commissions are made up of member states whose fishermen actually utilize the stock, and nonfishing states generally have little or no influence on management decisions. If, however, a strong concept of common heritage emerges, the nonfishing states may wish to participate more directly in the management of stocks than they do now.

States may also wish to share more inclusively in benefits derived from research and technological innovations. This raises certain difficulties, however. Under the conditions of open access, the sharing of research results and technological innovations is likely to lead to increases in congestion and economic waste. For example, it has been demonstrated that satellites could significantly reduce the costs of locating schools of tuna.[14] But this information, if shared widely among tuna-fishing states, would stimulate a race to the location and attract excessive amounts of vessels to a small area.

Finally, the knowledge that there are activities detrimental to the resources and the environment suggests that states may want to share in the prevention of damages to the common heritage. States may perceive the potential damages to be very great, as in the possibility of the extinction of a particular stock such as the Blue Whales of the Antarctic, or as in the possibility of irreversible modifications of the environment through the introduction of organicides into marine waters and the atmosphere. Even though a state may not anticipate utilizing the resource itself, it may view damage as a threat to the common wealth of the world and may wish to share in the prevention of loss. The degree to which this occurs and the techniques and institutions that might be adopted will help to characterize the definition of common heritage.

SUMMARY AND CONCLUSIONS

A conventional legal approach to the definition of the concept of common heritage has certain inherent difficulties. This is in part because some of the precedents, for instance the principle of the freedom of the seas, are of questionable validity as guides for future decisions on marine resources. And it is in part because an acceptable definition of common heritage cannot evolve gradually from past and present trends in customary and conventional law. On the contrary, at least with respect to fisheries and deep-seabed minerals, the development of the concept of common heritage as a meaningful term will require a marked change in the institutions and the law of the sea.

This conclusion has emerged from a functional approach to the definition of the term. The approach has examined the patterns of distribution in fisheries and the opportunities for more inclusive sharing under present trends, on the one hand, and under new concepts and arrangements, on the other hand. A thorough and comprehensive examination of common heritage should deal with all items of value that are sought from the seas.[15] This chapter, however, has been necessarily limited to a concentration on the values sought from fisheries, and particularly those that are economic in nature. It is recognized that other values, such as those found in international commerce and national security, will have an influence on a state's definition of common heritage. Nevertheless, an examination of the concept with respect to fisheries is important in and of itself.

One of the first steps to be taken is that of ensuring that perceptions of wealth are as realistic as possible. Perceptions of wealth in fisheries may range from the view that they are of negligible importance to the view that they are vastly abundant. Those that hold the former view will discount fisheries as an item of common heritage and may be willing to sacrifice any gains they might get from fisheries in order to achieve gains in other values. Those that hold the latter view will tend to define common heritage in terms of the widest free access. For if the resources are held to be so abundant that all can win and none will lose, then the production of wealth is directly related to, and increases proportionately with, the amount of effort and investment. Under this concept, the resources themselves have no value, for no one would be willing to pay for a resource when other similar resources can be gained freely by moving to a different spot. "A familiar formulation of overriding objective is that the great bulk of the oceans of the world should be maintained as a common resource, freely open to all peoples upon a basis of complete equality in the cooperative pursuit of the greatest possible production and sharing of values."[16]

It has been shown, however, that neither extreme is an accurate perception of wealth in fisheries. There are considerable economic gains that can be acquired, but they are not measured by the mass of undifferentiated protein material that lives in the sea. Man's efforts focus on a few dozen of the thousands of species, and these are found in only a few parts of the ocean. Catch by one fisherman can significantly reduce the catch available to others, and thus the resource itself becomes an item of value.

Thus there is both a trend toward and an economic need for exclusive access. The problem is that under present interpretations of the law of the sea exclusive access is associated with exclusive appropriation of wealth: that only those that exercise or acquire the right to fish can share in the distribution of wealth. This is supported by customary and conventional law: the principle of the freedom of the seas, the terms of most international fishery treaties (except for North Pacific fur seals), the Geneva Conventions, and the actions of individual states.

Certain possibilities for reversing the trends toward exclusive access were investigated. One of these would be to enforce open access through various measures. But, even if this could succeed politically, it would be extremely detrimental to the production of economic wealth. Each of the other possibilities, with respect to economic wealth in fisheries, requires the closing of free and open access, the extraction of some form of economic rent, and the sharing, in some manner, of the rent extracted. This calls for a dramatically different set of principles than those which have guided the use of the seas for the past three centuries. There appears to be no middle ground, no way in which past principles can be adapted to provide for a more inclusive sharing in wealth.

Thus, if the world community wishes to give meaning to the concept of common heritage in fisheries, it must take major steps in the adoption of new institutions and laws. Free and open access must be closed and those that obtain the privilege of fishing must pay for this privilege.

In order to make a wise decision, it is imperative that the means for achieving a wider distribution of wealth be characterized far more fully than they have in this paper. And this requires a significant increase in social science research. Among other tasks, studies should be made of the means for closing access, of the means for extracting economic rent, and of alternative systems and criteria for the distribution of rent. It is hoped that such research will get under way as quickly as possible so that states may be able to consider the alternatives when the conventions on the law of the sea are re-opened for consideration.

ARNOLD KÜNZLI

13 Science as the Social Property of Mankind

The question before us today is whether the triumphant procession of Western science and her legitimate child, technology, will not prove to be a Pyrrhic victory for mankind. As science took the the place of the old myths, and then of religion, it also took over their function. Science itself became a myth and a religion. We expect from science not only the overcoming of worldly need, but secretly also salvation. Even though the moon-landing was determined by political and, above all, military considerations, unconsciously the great masses were looking for God on the moon. Seen in this way the glorious conquest of the moon was a catastrophic defeat. Instead of the All we have found the Nothing. Science and technology have certified the unreliability of man relying on science and technology alone. The moon has denied Francis Bacon in proving him to be right. It has proved that knowledge is in fact power and that it gives us a power over nature which can realize what until recently seemed unrealizable. But at the same time the secret longing of that will to power has been cruelly disappointed, expecting that it would, with the help of science, succeed in discovering the hidden sense of this world.

The result of the moon-landing experiment is the recognition of the senselessness of a knowledge that expects redemption from its own power. When the first man stepped on the moon, the age of enlightenment which elevated reason to a religion came to an

---

ARNOLD KÜNZLI, author and journalist, is lecturer for political philosophy at the University of Basel.

end. Therefore, what we need today is an enlightenment on enlightenment.

The same may be said about another belief which has inspired knowledge since the days of Socrates: the belief that knowledge necessarily produces virtue and so a good life and also a good society. Here Nietzsche saw more deeply when he wrote: "There is no pre-established harmony between the furtherance of truth and the welfare of society." Statistics on the number of German professors who became Nazis, for example, would confirm Nietzsche's thesis scientifically. Napalm is also a product of scientific knowledge. Since Max Scheler differentiated between a knowledge of culture, a knowledge of redemption, and a knowledge of domination and efficiency, we must realize today that a knowledge of culture does not save us from political folly or even political crime, that a knowledge of redemption has experienced a lunar dementi, and that a knowledge of domination and efficiency can serve and does serve on a large scale to set up a system of knowledge which is oriented to success and profit in the service of an unlimited struggle for power over nature and man. (The struggle for knowledge which Aristotle, in the first sentence of his metaphysics, called a struggle belonging to the nature of all men, is in almost all of its forms a struggle for domination and power.)

With that I am not professing a pessimism over culture or civilization, I am only asking the old question about the relations between knowledge and values in a new form. The belief that science in itself is one of the highest values producing in a somewhat self creative way permanently new values is shaken today. Science as such is value-neutral. Science produces simultaneously means of destroying and means of healing life. Its left hand does not know what its right hand does. But science is not pursued in a vacuum, but rather in societies with certain socioeconomic and political structures reflecting their interests and values. The consequence is, on the one hand, that science in its premises—for instance, regarding the object-selection—is determined, or at least influenced to a greater or lesser extent, by social forces. In a time in which scientific research becomes always more costly, money and the private or state-interest behind it are playing an always greater role in the object-selection; witness the predominant importance of the Pentagon in the distribution of research projects. In the Soviet Union it will hardly be otherwise. The results of scientific research are to a large extent being utilized for the realization of the aims of the people who give the orders and the powers determining political and economic direction as a whole. So science, although as such value-neutral and only interested in enlarging its area of knowledge-power, is taking over—with reference to the object-selection and the utilization of the research results—a social function which is largely determined by the interests and normative axioms of those who in a given society exercise the real power. That science, which for Theodor Geiger has the cultural-political function of being the bad conscience of power, has become an instrument which can be used for good or for evil.

But there is another problem. In the course of the scientific-technological revolution, science and technology developed autonomous power, thanks to spectacular success in overcoming limits which until now nature has set for men and in eliminating so many of the burdens and sufferings imposed until now on men by nature and society. This power entered into a competition with the powers of particular interests and politics. In using science and technology more and more for the realization of its aims, politics has also become more and more dependent upon them and in danger of abdicating its own practical, value-oriented powers. More and more political problems are regarded only as purely scientific and technical ones. This leads to a curious contradiction which seems to me to characterize our actual political situation to a large extent. On the one hand, science and technology have become important instruments of politics and of the interests and established principles behind it. On the other hand, science and technology have developed a life of their own which has become a new power factor in society and with which the political process must reckon. Today a scientific discovery can have revolutionary consequences for politics. We need only to think of the atom bomb and the constant changes in the field of arms technology. But both phenomena, science and technology as instruments and as a competitor of the political process, bring politics into increasing dependence upon a power in itself neutral to values. And that means: politics is becoming in the real sense of the word "de-valued."

If now we bring economics into relation with our observations, it follows—as Marx already saw in his *Grundrisse*—that science and technology have become the most important productive force today. They are not only decisive productive forces but also a no less important means of production. This fact leads us to our proper theme, namely, the question of the ordering power that stands over these forces and means of production which are—think of the nuclear arms and of the signs of ecological catastrophe—decisive even for the survival of mankind as such. In the so-called West this power of disposal lies in the hands of an increasingly anonymous private economy and the government with its administration. The latter represents indeed theoretically the will of the people and is subordinated to the control of parliament and public opinion, but we all know how far practice has withdrawn from the original theory of parliamentary government. Confronted with what the late President Eisenhower called the "military-industrial complex," the public and its democratic representatives are to a large extent powerless. In the so-called East, the situation is partly better, partly worse. It is better in so far as the ordering power over science and technology does not lie in the hands of profit-oriented private enterprises with their special interests. The situation is worse in so far as a small "elite" of party and state, not freely elected and not democratically controlled, representing essentially its own interests, with a torpid and sometimes even corruptible bureaucracy, has control of the forces and means of production. As far as Europe is concerned, Yugoslavia is the exception, but even here also the practice has some difficulties keeping in step with the theory.

So we have to face the fact that science and technology, the decisive productive forces and means of production which today are in a position either to ameliorate radically the situation of mankind and of each individual or to destroy mankind as such, are subordinated, in the West as in the East, to the power of disposal by largely anonymous, uncontrollable, particular "power-elites" with corresponding interests. At best they may in some spheres lead a life of their own which more than ever cannot be subordinated to social control. How perilous the situation is may be seen in the ecological crisis, which is a result of the social unreason and irresponsibility of a science and technology oriented essentially to particular interests. This is also demonstrated by the documents published by the Swedish Institute for Conflict and Peace-Research in Stockholm, revealing that since 1945 mankind has been on the brink of a nuclear catastrophe at least sixty times, owing partly to technical failures.

In my opinion one can draw only one conclusion from this situation: namely, that we hang suspended in potential suicide as long as we allow particular groups or interests to determine decisively the object-selection, the direction, and utilization of science and technology and to exercise control over them as forces and means of production.

But what would be the alternative? I see the only alternative—as utopian as it may appear to many—in conferring on science as the theoretical foundation of technology the status of a social property of mankind. The leading idea—apart from the dangers mentioned—is, that science is a heritage transmitted to us from past generations. It has been effective in times in which there were no private enterprises and no party-elites, indeed not even national states with ministries of defense and bureaucracies, so that already from a historical perspective it seems senseless to allow this heritage to remain under the controlling power of such limited forces.

But another thought seems to me to be even more important; namely, the consideration that science, and far more technology, has become a collective enterprise in which besides the administrative organization, financing powers, and members of the research team, a great number of persons is participating, including the workers who dig out the uranium for the experiments of nuclear physics and the employees of electric and power companies which serve the laboratories. With their contributions to scientific research these employees have, as have the scientists themselves, acquired a right of participation in the direction and control of the enterprise. In addition to these more or less direct participants in the scientific process, there are many indirect participants, namely, society as such. This joint social participation has several aspects: it takes, for instance, the form of tax-money which the state is investing in scientific projects, institutions, universities, libraries, and so on. Or it may take the form of basic construction and maintenance and service, also financed by taxes and indispensable for scientific research and accomplishment: one has only to mention highways, waterways, waste-removal, and so on. Furthermore, it takes the form of social costs imposed on society by private as well as by state enterprises. I would only mention the intervention of science and technology in nature, with consequences for the whole society, and the pollution of our natural environment. But social problems arise also when a great scientific-technological project is abandoned for one reason or another. On the whole, science and technology are today so closely connected with society that to society must be conceded the right to regard them as its property.

Now I realize only too well what problems and difficulties emerge if one tries to define concretely what sort of form science as social property shall take. The aim of these thoughts is only to present the idea for discussion. The answer to such a question could hardly be given by one person alone, but would rather be the task of interdisciplinary and international study-groups. Here I shall confine myself to sketching some postulates.

Perhaps the decisive question is the one concerning object-selection. A way should be found and institutionalized to permit society—at least for greater projects—to have an important part in determining the object-selection, maintaining at the same time an optimum of freedom in scientific research. As for this freedom one should say that today, in the West and in the East, it is partly a myth, because too often it is only a monopoly of professors, directors of institutes or leaders of research-projects, who themselves are dependent upon people who give them the orders or the money. A democratic social co-determination of object-selection would, therefore, in my opinion, not limit but rather enlarge the sphere of freedom in scientific research.

The second problem is that of the tendency. As a result of its entanglement with society, science inclines always to reflect the tendency and the contradictions of the society in which it is pursued. In our society of performance and success—to mention only these—science is oriented essentially toward performance and "success." These represent in some way the immanent requirements which science tries to satisfy. So one could speak of the self-satisfaction of science. However, if science were declared a social property, a permanent discussion between science and society about the legitimate requirements of society would be institutionalized, directing the tendency of science from self-satisfaction, or from the satisfaction of the desires of particular interests to the satisfaction of legitimate requirements of society, and developing a new conception of scientific progress.

A third problem is that of control. If science is declared a social property society has the right and the duty to control, ensuring that this social property is not misused for purposes opposed to the legitimate requirements of society. This control must also be a social control. It must not be allowed to fall into the hands of a national or international bureaucracy or even a party apparatus. And that means: it must be a social self-control, in which science is participating in a measurable way, again to secure an optimum of freedom of scientific research.

Of utmost importance for the functioning of a social self-management of science is finally the old liberal principle of publicity in the twofold meaning of the conception: on the one hand, the process of object-selection, tendency-discussion, and self-control must be transparent and must take place in public; on the other hand, with the new rights the public is given new responsibilities to which it can rise sufficiently only if it is adequately informed about the social function of science and if in this process of information it is becoming conscious of its new responsibility. To mention an extreme example: if a research-project aims at finding a new means of mass-annihilation, or if an industrial enterprise is producing napalm, then the last assistant worker and the last secretary co-working on this project, and beyond that, society as such becomes responsible in the same way as the people who give the orders or the money, or the scientists and the directors of the project. Social property implies social responsibility.

Of course, the concept of science as social property cannot be considered in isolation. One can hardly declare science alone a social property, so long as society as such is still governed by the principle of private economy with state-intervention or state-economy with party-intervention. The concept touches the roots of the social systems in West and East. Its leading idea is the emancipation and maturity of man as a social being. But if, on the other hand, we wait for a revolutionary change of our social systems before risking the change of our scientific system it would, perhaps, be too late for the latter. Someone should act as a pioneer. It seems to me that no one is called upon to serve in this pioneering role so much as science, for science in the first place is responsible for raising the question concerning the survival of mankind. And it is animated, or at least should be, by the critical mind which is a precondition for every authentic change.

It has been stressed, especially from the side of the scientists, that the questions here presented are ultimately philosophical ones. They concern the relationship between man and nature, individual and society, knowledge and interests, responsibility and history, and above all the question of the sense, the values and the norms of all our thought, study and action. It seems to me that philosophy has in this connection a new task and a new function. Philosophy, too, is a social property, and even in its most esoteric form it remains connected with the society in which a philosopher lives. What the philosopher can do in this period of transition or preparation, is to formulate the problems, to trace the aims, to stimulate the discussions, to put critically the question of truth, to confront the special requirements of the experts with the general requirements of society, to question all knowledge of its possible interests, to put the actual in the context of history, to ask inexorably the question of "why?" and "for what?"; in other words, the question of normative values, and to try to achieve a consensus about the values which can make our life worthwhile.

STUART A. SCHEINGOLD

14 Lessons of the European Community for an Ocean Regime

In its most general form the lesson of the European Community is in its capacity as a functional regime to concert national policy on matters of real economic importance. The term functional here refers to a territorial regime which has specifically defined tasks and authority. Accordingly, there is reason to believe that the experience of the European Community may be applicable to the proposed ocean regime. This introductory section defines the characteristics of a functional regime more carefully, and is followed by an exploration of conditions that seem to have been necessary, if not sufficient, for the success of the European Community.

A functional regime may serve either a great many or only a few purposes. The European Community as it is presently constituted impinges upon virtually all sectors of the economy, but it began in 1953 with the pooling of just coal and steel. Today it includes the Common Market, Euratom, and the Coal and Steel Community. Similarly, the authority delegated to the institutions of any functional regime may vary as widely. Within the European Community it is possible to distinguish two fairly distinct patterns.[1] In certain sectors, agriculture and transport for instance, the Treaty commits member states to do no more than *seek* a joint policy, and approval by these states (in the Council of Ministers) is a precondition to action. Once policies are established, how-

---

STUART A. SCHEINGOLD is a Professor in the Department of Political Science, University of Washington, Seattle.

ever, or if Treaty obligations are explicit—as is the case with tariffs, for example—the Commission may implement them with binding decisions subject to review only by the Community's Court of Justice.

What is constant in these variations on the functional theme is that the commitments of the member states can be specified at the outset, and that major increases in scope and authority of the regime are dependent both officially and factually on the agreement of these states.

Similarly, one can say that functional regimes seem able to transcend territorial boundaries. That is to say that the jurisdiction of functional regimes, unlike that of traditional states, is defined by task rather than geography. Within the boundaries of the European Community, the national and regional systems share the authority to deal with a broad range of economic problems. Of course, it is clear that an ocean regime would probably be more radically nonterritorial than the European Community.

But insofar as this implies intrusions into the established boundaries of the coastal states, the experience of the European Community suggests that suitable arrangements can be worked out. With respect to the "public domain" of the sea, one can only assume that a totally functional regime would be more suitable precisely because it would escape the confines of subdivisions which were not related to its tasks.

With the defining characteristics of functional regimes specified and a *prima facie* case made for the suitability of this classification to an ocean regime, one can try to distill from the experience of the European Community the requisites of success. The primary emphasis is on those institutional arrangements and processes which are conducive to sustaining and stabilizing the integrative process. Research on the European Community comes closer to yielding general principles of systems maintenance than to revealing how and why such regimes are initiated. In particular, it seems clear that with respect to scope and authority, functional regimes must be consensual rather than coercive, and this in turn suggests certain standards for evaluating institutions and processes of the European Community system. It is much more difficult to generalize on the basis of the European experience about what specific conditions are conducive to establishment of a functional regime.

THE INITIATION OF THE EUROPEAN COMMUNITY

The European Community was created out of a widely shared sense of common crisis. At the close of World War II much of Europe lay in ruins; the political systems of the individual states were in a shambles; and confidence in the nation-state as a source of security, welfare, and democratic values was, to say the least, badly shaken. In addition, control of those countries eventually joined together in the European Community—France, Germany, Italy, and the Benelux countries—tended to be in the hands of moderate parties of the center—in particular Christian Democrats—that shared a faith in democratic pluralism and welfare capitalism. What this meant was that beyond common problems was a kind of inchoate consensus on the values which should guide the choice of solutions. Moreover, the Christian Democrats tended to favor transnational approaches.[2] Finally, there was strong, although far from unanimous, support in *economic* theory for both the unifying and the wealth-enhancing effects of economic integration—whether in the form of a customs union, of a common market, or of a full economic union. The promise of economic integration was expansive not defensive: economic integration was not simply a way to resolve a crisis but promised a bright future.[3]

With all this common ground, it might come as a surprise to find out that the European Community was not established by spontaneous acclamation, but through a process of hard bargaining. The Community is a tribute to the resourceful manipulation of converging interests and is based on a series of compromises which some would characterize as fundamental and ultimately debilitating contradictions.[4] The bargaining trade-offs that supported the constitutive compromise were peculiar to postwar Europe and are unlikely to have any bearing, as such, on the establishment of an ocean regime. On a more general level there may, however, be some instructive lessons to be learned, although I shall leave that judgment to those more familiar with the Ocean Regime.

Most obviously, there is some reason to associate initiation with a widely shared sense of present or impending crisis.[5] Second, it is probably true that the contribution of the functional regime to handling or averting the crisis must be relatively clear and the deficiencies of nation-states acting individually equally apparent.

Clearly, the conditions for joint action in the oceans are not so propitious as they were in postwar Europe. The ideological divisions will make agreement on means if not necessarily on ends very difficult. The absence of pluralist politics in many countries may well make it more difficult to discover, disaggregate, and recombine interests into acceptable bargaining packages. And while the shadow of ecological crisis hovers over us all, it is not so salient, so apparent, or so mutually perceived as was the European crisis at the close of World War II. Still, there are some significant grounds for comparison with the European situation.

THE CONSENSUAL REGIME: EFFECTIVE POLICY MAKING

A sector-by-sector consideration of the European Community indicates that it has done some things well and other things badly. Under the right conditions, however, its functional regime is capable of making and implementing policy even in sensitive economic sectors which have an important impact upon the lives of many citizens and influentital elites. The purpose of this section is to consider the political and legal processes associated with effective policy making. Insofar as possible, I shall avoid a discussion of the European Community, as such, and focus on those matters which seem to transcend the Common Market and European contexts.

THE POLITICAL DYNAMICS OF EFFECTIVE POLICY MAKING

The participants in the initial Coal and Steel Community discussions had committed themselves even before the first meetings to two preconditions suggested by Jean Monnet: (1) pooling of their coal and steel industries rather than free trade in the two commodities and (2) an organization headed by an agency with the authority to make binding decisions. Above and beyond the economic justification for these preconditions, they were seen as the necessary minimum for an effective regime. Free trade arrangements were too ephemeral: what was needed was the kind of industrial interpenetration which would inextricably link these basic industries of the member states.[6] Strong institutions were viewed as prerequisite to effective implementation of any program agreed upon, and thus as the way to avoid the paralysis which often overcomes international organizations whose executive agencies can act only when all the member states agree. The Community has of course endured and its institutions have proven themselves capable of effective action, but these achievements are not explicable entirely in terms of the initial rationale.

The success of the Community's institutions is not due so much to their power to coerce compliance as to their capacities for generation of a political forum.

How do we explain this apparent anomaly? It has turned out that the secret to effective action is not coercion but initiative and, to a lesser extent, attracting and satisfying a constituency of relevant elites. It has become increasingly clear that executive agencies cannot act effectively in the face of opposition from the member governments acting in concert, regardless of their formal authority. On the other hand, an enterprising and astute executive has the resources to mold consensus that is more than just a minimum common denominator of the positions of the individual member states. Perhaps the chief among these resources is the executive's understanding not of the merely acceptable, but of the member states' priorities and urgent needs.

Of course, the success of such bargaining is also directly related to the nature of the policies and programs being considered. Generally speaking, it is easier to negotiate successfully when the matter under consideration is of strong importance to one or more nations or to influential elites within the member countries. More particularly, it is easier to form a consensus when joint policy-making promises expandable opportunities.

This once again underscores the importance of Monnet's determination to move beyond free trade and at the same time to cast some doubt on the initial justification for the pooling of coal and steel. There does not seem to be much evidence that the coal and steel industries of the six member states are inextricably intertwined. On the other hand, the *decision-making processes* have become markedly—if not inextricably—intertwined. It can, moreover, be persuasively argued that this stems in part from the importance of the decisions entrusted to the Coal and Steel Community. That is to say, it was the willingness to go beyond free trade that laid the basis for a real political forum, which seems in turn to be the necessary pre-condition to effective joint policy making.

THE LEGAL DYNAMICS OF EFFECTIVE POLICY MAKING

The Community legal system departs from normal national and international legal processes in a number of ways that relate directly to the notion of a consensual regime capable of engaging influential elites and the member governments in a process of joint decision-making. The features I have in mind make the Community system unusual, if not altogether unique among international organizations. Three innovations are particularly relevant: (1) community law takes precedence over national legislation, (2) community enactments have the force of law in the member states, and (3) community legal questions arising in national litigation may be referred to the Court of Justice in Luxembourg. At first glance these departures may seem more likely to promote the independence of the regional system and the binding authority of its institutions than to serve as an agent of consensual decision-making patterns. My contention is, however, that while in form the keynote may be autonomy and constitutional coercion, in practice the cause of interpenetration and consensus is served in a manner analogous to the political processes already discussed.

To put this all in somewhat more familiar terms, I want to pose some questions about the desirability of assessing the Community in terms of standards provided by a *federal* model. I would consider the legal attributes of the functional regime without reference to some brighter future but with respect to the effectiveness and stability of the present system. This orientation is chosen because it relates better to the purposes of this paper, and also because I am inclined to think of the functional regime in quasi-permanent rather than transitional terms.

What the *direct applicability* of Community law means is that Community rule gives rise to individual rights and obligations within the member states. That is to say, the obligations created by the Treaty are binding not only upon the member states as such—which in itself is rare among international organizations—but individuals who fail to fulfill obligations under the Community Treaty are also subject to enforcement actions by Community institutions and ultimately answerable to the Court of Justice of the European Community. Conversely, remedies for violations of individual rights under Community law include litigation before national or Community tribunals. This aspect of Community law was initially set forth by the Court of Justice in 1963:

> The Community constitutes a novel juridical order of international legal character for the benefit of which the states, though only in limited areas, have limited their sovereign rights and the subjects of which are not only the member states but also their national; consequently, Community law, independent of the legislation of the member states, creates not only burdens upon the individuals as such but, conversely, is also apt to entail rights which enter into their legal patrimony.[7]

The *supremacy* of Community law was asserted by the Court of Justice most memorably in a 1964 case which called into question the nationalization of electric power in Italy.

> The preeminence of Community law is confirmed by Article 189, under which regulations are "binding" and "directly applicable in each Member State." This provision, which contains no reservations, would be meaningless if a Member State could unilaterally nullify its effects through a legislative act that could be asserted against the Community texts.
>
> . . . The transfer by the State from their internal legal systems over to the Community legal order, of rights and obligations to reflect those set forth in the Treaty, therefore entails a definitive limitation

> of their sovereign rights, against which a subsequent unilateral act that would be incompatible with the Community cannot be asserted.[8]

The point to be underscored about the supremacy of Community law is that, taken together with *interlocutory* procedures provided for in Article 177 of the Common Market Treaty, it significantly enhances the possibilities for interpenetration. This position will be developed below, but its crux is that parties can subject national rules to Community standards in litigation before national courts.

Direct applicability, supremacy, and even interlocutory proceedings seem to point up the hierarchical character of the Community system and its capacity to utilize constitutional coercion against the member states. Certainly, they could be utilized in this fashion, but generally speaking consensus has been the style of the European Community. There have been flirtations with constitutional coercion, particularly within the Coal and Steel Community. The High Authority, unwilling or unable to generate sufficient political support for its policies, attempted to use litigation to bring recalcitrant member states into line.[9] This approach was never really successful and is, in any case, not characteristic of the consensual manner in which the Community normally solves its problems. The outlines of this consensual system were considered above, and the question dealt with here is how the judicial process fits into them.

Clearly, litigation is part and parcel of the Community system; Community problems are regularly adjudicated in both national courts and in the Court of Justice. Moreover, much of this litigation is a direct function of the applicability of Community law within the member states. It is probably also related to the supremacy and to the binding character of Community law. On the other hand, basic disagreements among the member states—even when they involve obvious violations of the Treaty, as in de Gaulle's partial boycott of Community institutions—are not settled by litigation.

> People have regretted that in last year's differences between France and its partners, the legal implications of the French absence in the

> Council was never submitted to the Court. That is much too legalistic an approach; it would have been the definite end of the communities as communities if the opposing parties had gone to law and asked a ruling on details that were no more than the juridical top of a political iceberg.[10]

It might even be said that governments are never forced to act contrary to what they perceive to be their vital interests, and that, in any case, such disputes are resolved by negotiations which focus directly on these vital interests.

What then is the role of litigation? Does it serve any enforcement functions? Actually enforcement litigation against governments which resist the application of a given rule is quite common. While there are any number of reasons for such resistance, resolution by litigation ordinarily suggests that the governments, as such, were not in basic opposition to the policy. More than likely it was a disagreement about the way in which the policy was being implemented—largely a problem of interpretation, in other words. Moreover, while the cabinet may have given formal authorization to resist or to litigate, it is unlikely in such instances that resistance *per se* was really a matter of government policy. Thus, litigation and the legal process in general serve primarily such goals as coordination, uniformity, and communication, rather than policy making. Conflicts are resolved by litigation, but this probably does not include conflicts over first principles.

This is not to say that only unimportant or minor questions are dealt with by litigation. The kinds of problems which come before the Court of Justice and national courts relating to agriculture, the customs union, or antitrust law often have a significant impact on large numbers of people and may be deemed vital by Community administrators and even by some national officials. Moreover, many decisions of national and community courts relate to basic kinds of structural questions and help to establish the patterns into which conflicts can be channeled. But the creation and development of these patterns is more important than whether they vindicate the prerogatives of the Community. The legal process in the Community is not utilized to control the member governments so much as it is utilized to engage them. Litigation takes place within consensus and may serve to clarify the details of that consensus, but litigation is not an effective means for forming or imposing consensus.

THE STABILITY OF FUNCTIONAL REGIMES

It has often been said that the European Community must either expand or wither, but the evidence to date suggests that this is not true. The Community shows distinct signs of stabilizing as a partial economic union with institutions that fall somewhere near the mid-point of a continuum between a traditional international and a federal regime. The European Community has, therefore, been a distinct disappointment to some of its most enthusiastic supporters, who are inclined to characterize these equilibrium tendencies as self-encapsulation. They are pleased that the Community has been able to accrue sufficient authority to accomplish most of its more precise goals but distinctly unhappy at the inability of the system to generate sufficient support to break out of its economic confines and spill over into diplomacy and military security or, for that matter, to establish institutions with a more federal cast to them.

But from the perspective of an ocean regime equilibrium would seem to be more an advantage than a drawback. In other words, the message of the European Community is that a functional regime can operate effectively within a prescribed area. Moreover, it would simply be incorrect to identify stability with stagnation, since measured by almost any standards the system has grown. The limits of that growth have, however, been defined by the tasks initially specified in the Treaty. In other words, the system created by the member states remains in a very basic way responsive to their shared purposes and converging interests.

In looking a little more carefully at the nature and limits of growth in functional systems, the most readily observable signs of growth in the European Community are: (1) the increased number of sectors subject to joint decision-making and (2) a tendency for the locus of decision-making to shift towards Community institutions. The first point is clear and needs little elaboration beyond recalling the increase in scope from coal and steel to a broadening range of activities connected with the establishment of the Common Market. The concept of locus relates to the way in which authority for joint decisions is shared between Community and national institutions. What has happened over the years is what might be termed a shift in the center of gravity with Community institutions playing a larger role in many matters of joint concern. This is most obviously the case in agriculture but it is true to a lesser extent in social welfare policies, countercyclical policy, and elsewhere.[11]

To a large extent, this growth has been in response to problems created by the initial integrative steps. Economic sectors are interdependent, so a joint solution by the member states to problems in one sector tends to release pressures for joint problem solving elsewhere in the economy. Initially, it was believed that it would be through such an economically determined chain reaction that a complete economic union would be welded and that this in turn would lead to a single European polity.

This determinist logic has now been largely abandoned. Sectoral integration does indeed release such expansive pressures, but there are also strong counterpressures which underscore the autonomy of economic sectors. These pressures manifest themselves in positions advanced by governments and influential interest groups, and these positions are only in part determined by the economic calculations of gain and loss. It is then in the clash of interests within the Community political arena that growth is determined, rather than as a direct and predetermined consequence of economic forces.

The significant but limited development of the Community is also reflected in the response of Europeans to integration, and once again we find the European Community disappointing some of its most ardent fans. It was at one time believed that economic integration could catalyze a shift in loyalties from the nation states to the Community in a kind to two-step process. First, it was hypothesized that interest groups would tend to shift their attention to the new institutions as it became clear that more and more decisions were being made at the European level. A re-direction of loyalties was expected to follow, once the rank and file of these groups realized that the bread and butter problems were no longer being resolved by the member states.[12]

So far as we can tell, this does not seem to be happening. It is true that the levels of support for the projects of the European Community have grown rather steadily among elites and mass publics, but this does not seem to be at the expense of national loyalties and, in addition, the working class segments of the population seem to be resistant to the appeals of European integration. Moreover, the activities of the Community tend not to impinge directly on the lives of the mass public since policy implementation is most often through national authorities who, for example, are responsible for administering those regulations affecting trade among the member states and with other countries. Finally, to the extent that the Community regulates rather than provides goods and services, the perception of impact by individual citizens is likely to be diminished.[13]

Another related matter is that the Community does not seem to have a single constituency but is composed of a number of constituencies that tend to mobilize individually when policy matters relating to them are being dealt with but are otherwise somewhat passive. Thus, the institutions of the Community do not really aggregate an increasingly large coalition of supporters who mobilize on behalf of programs merely because they advance the cause of integration. Once the demands of a given group have been met and a satisfactory policy established, it is not at all uncommon for this group to resist further growth in order to protect the gains already achieved. Thus, growth tends to generate—simultaneously with stimulants to growth—conservative pressures which are likely to inhibit future growth.

If it is difficult to expand within the economic realm because of the autonomy of functional contexts, it is even more difficult to extend joint policy-making to military and diplomatic matters. It has been persuasively argued that decisions involving matters of national security are not as amenable to bargain and compromise as are arrangements of material matters that can be easily subdivided.[14] In addition, as a direct consequence of the broad reach of the Common Market and the interdependence of economic problems, most economic elites are eventually drawn into the Community's political arena, and once their needs and priorities are understood, the possibilities for constructing a package deal that will attract their support is increased. There is in fact some evidence that participants in the process are "socialized" by the experience and more easily brought into coalitions. Military and diplomatic elites—particularly the former—stand quite apart from this process, and Community officials are not in a position to understand their needs and priorities and, in any case, probably do not have access to the kinds of rewards that would properly bait the bargaining hook.

The first conclusion that might be drawn is that functionally specific systems may have certain "natural boundaries." While these boundaries can be opened and the system extended, this will probably not be the result of forces generated internally, but will require the same kind of constituent and consensual act by the participating states that established the initial project. It is difficult to specify just where these natural boundaries are, but task expansion seems directely related to intensity of relationship and mutuality of concerns among those political constituencies affected by what has already been undertaken and what is proposed.

In the case of the European Community these natural boundaries seem to lie between economic and military matters. This would not necessarily be true of an ocean regime if military needs and the exploitation of resources impinge upon one another. The European Community faced just such a conflict in an atomic energy project which came close to foundering in large measure because of the impossibility of extricating problems of generating nuclear power for peaceful uses from France's determination to go ahead with its nuclear military program.

Within its natural boundaries the functional regime does generate certain growth pressures. At the same time the conservative counterpressures which tend to inhibit growth are also likely to be supportive of attempts to rescue the system in crisis situations, since these elites obviously have ample stakes in the continuation of the system. The results, as I see it, are strong equilibrium tendencies in functional regimes built upon the very narrow base of a single test case, the European Community.

But the effectiveness and stability of functional regimes is not in some mysterious way foreordained. Whether an ocean regime is launched and whether it prospers depends, in large measure, on the nature of the joint concerns and the individual interests that such a regime would exploit, as well as on the quality of the leadership that can be mobilized on its behalf.[15] In other words, a final answer to the questions raised in this chapter must be based on a detailed analysis of the real and perceived stakes [16] nations may have in an ocean regime.

CLAIBORNE PELL

15 The Political Dimensions of an Ocean Regime

Nineteen sixty-nine was not a vintage year in the United Nations for negotiations on the uses of the seabed. I must add that this was true not only in the United Nations; it was not a vintage year for the making of any real progress in high-level decisions on the matter within the United States government. My hope is that this new decade will open a more benign and decisive era in this and in other areas of unresolved conflict. Views that emerge from these meetings of Pacem in Maribus will not necessarily be reflected in the policies of governments. But expert knowledge applied to consideration of possible alternatives is a prime source of information and ideas, and so necessarily of influence.

My interest is not that of an expert in any single aspect of this complex undertaking, but that of a politician who has out of deep concern become an expert generalist. A politician's great limitation is that he must deal in the art of the possible, and it is under that handicap that I take part in the effort to achieve progress in an area where, some say, one must, like the man of La Mancha, dare "to dream the impossible dream, to fight the unbeatable foe, to bear with unbearable sorrow, to run where the brave dare not go." The necessities of compromise are seldom the stuff of dreams.

Nevertheless I believe that Ambassador Pardo's initiative in

United States Senator CLAIBORNE PELL (Democrat, Rhode Island) is the author of Senate Resolution 33 which he introduced in the 91st Congress and which presents one of the most fully developed models for a treaty establishing an international Ocean Regime.

proposing that the United Nations take up the problem and my own almost simultaneous introduction of a draft treaty proposal in the United States Senate have both done something to focus the conscious awareness of governments—if not the grander dreams—upon an impending source of conflict too easily lost in the welter of more immediately compelling crises.

Unfortunately, recognition of the problem does not constitute either a policy or a solution. Senate Resolution 33 makes some suggestions about both, and many participants in this meeting have testified at hearings on it. The appended current version of the resolution is not substantially different from the one I introduced in the United States Senate more than two years ago. This is not because I have not learned a great deal from my hearings or from United Nations and other meetings, but because I intended that both versions of the resolution spur accomplishment of purposes still unaccomplished: my interest is to stimulate my government into making the necessary decisions and to suggest reasonable and workable solutions to the two key issues: the boundary of national jurisdiction on the seabed and a regime for the seabed beyond that could provide—with minimum bureaucracy—the requisite elements of physical and financial security without which it is unlikely anyone will benefit from the riches of the seas.

Last December the General Assembly of the United Nations adopted several resolutions relating to the seabed. Two of these are of some significance in that they were passed by bare majorities which do not reflect the weight of actual interests. One calls upon the Secretary-General to consider the desirability of convening a conference on broad issues of the law of the sea, including questions of the territorial sea, the contiguous zone, and fishing zones, as well as the boundary of national jurisdiction on the seabed. A second resolution, the so-called moratorium resolution, states that, pending the establishment of an international regime, states and persons are "bound to refrain from all activities of exploitation of the resources of the area of the seabed beyond the limits of national jurisdiction."

These resolutions spotlight the degree to which the key issue of the boundary remains both troublesome within the purview of the U.N. seabed committee and an unresolved question in the policies of my government and those of some other nations whose interests in the seabed are substantial.

In the United States, the boundary question involves considerations of national security, of freedom of the seas, of the varied interests of the oil industry and of other industries who may ultimately be mining the deep seabed, all of which are to some degree conflicting. Thus the Department of State and the Administration are still, I regret to say, pursuing with vigor their "no policy" policy. In other nations the question is similarly enmeshed in a variety of considerations political, economic, and geographic.

In sum, in these two years we have acquired not only technical knowledge of the seabed, but greater knowledge of political considerations which bear on decisions as to its future. The time has come to seek solutions in terms of compromise and accommodation of realities that will not go away.

Among these realities, or obstacles in the view of some, are the interests of the United States, of the Soviet Union, and of other maritime powers, and these interests must be adequately recognized and protected if any regime is to exist in fact. I regret that some of the resolutions adopted in the last session of the General Assembly seemed to "endorse what are basically the fears of a small group of countries"—if I borrow Ambassador Pardo's very apt comment in the General Assembly debate.

I am particularly concerned about the so-called moratorium resolution. In the absence of any agreed definition of the limits of national jurisdiction such a resolution is illusory. Nonetheless it is an illusion which serves well as an excuse for those who would extend the limits to the maximum in order to avoid just such clouds of uncertainty.

The present polarization of views within the U.N. does not serve either the cause of orderly and efficient exploitation of the seabed or that of insuring some common international benefit. It thus becomes more imperative that we seek an agreed boundary and establish the guidelines for a regime for the ocean bed beyond, before those choices are dictated by technology.

Senate Resolution 33, I am still persuaded, offers the basis of reasonable compromise between conflicting interests both within the United States and among nations, and I here review briefly the considerations that underlie my proposals with regard to a boundary and with regard to a regime.

It has been charged that a few of my proposals about a regime are vague, as indeed they are—intentionally so. They are outlined more in terms of functions—what a regime should do—rather than how it should be structured. The World Bank and possibly some other existing international bodies offer useful patterns for combining technical expertise with adequate recognition of political and financial power. But before voting formulas and composition of executive bodies can usefully be dealt with, it is necessary to agree on what such a regime should do, and what it should not do, for those objectives dictate in part the nature of the regime.

In fact, the functions of the regime appear to be of first priority, since it becomes increasingly evident that the boundary issue is almost unmanageable without prior agreement on a regime which might lay to rest the small nations' fears of being shut out and the great nations' fears of being buffeted by the paper majorities of small nations.

Senate Resolution 33 addresses the requisites of an acceptable regime that would be flexible and adaptable to future changes in technology and to the growth of confidence in its capacity to act fairly. The major objectives of the regime proposed in my treaty are these:

(a) to provide physical and financial security for exploitation, including assured tenure for substantial terms of years;

(b) to establish rules governing safety, pollution, and other hazards to navigation and to the ecology of the sea to insure the peaceful adjustment of competing uses and to establish rules of legal liability for accident, injury, and criminal acts;

(c) to provide a means, through license fees or royalties, whereby funds might become available for agreed international purposes; and

(d) to provide a procedure for the resolution of disputes.

Clearly a system of registration alone would not encompass these functions and could give only an appearance of order where in fact none would exist. On the other hand, a system that could provide the requisite financial and physical incentive to encourage the enormous investment risk of exploitation precludes, in my opinion, any regime that would vest in the U.N. jurisdiction and responsibility for exploiting the riches of the deep-sea bed or any regime which would accrue for the U.N. all profits of such exploitation. The end result is likely to be an ineffective regime and little expectation of international revenues.

If the fears of small nations of losing out on this last great bonanza—if and when it should become one—are perhaps well founded in history, what is equally probable is that exploitation will not be stayed indefinitely by diplomatic argument once technology lends itself to economic production. Indeed, the practice of leasing vast areas of marine territory for exploration for and exploitation of oil is growing by leaps and bounds both in area and in distance from the shore.

It was my hope and intention that proposals for a regime responsive to the realities of maritime and technological power and insuring both financial incentive and a means of obtaining revenues for agreed purposes would make it easier for nations to settle on a relatively narrow reach of national jurisdiction on the seabed. Lewis Alexander in his very useful paper outlines the various possible formulas separately and in combination. I am not wholly in agreement with his conclusions but I do believe that the pluses and minuses he has detailed for the various proposals are extremely well thought out.

He thinks more favorably of the existing Shelf Convention than I, although an effort to refine that Convention by an agreement on definitions would perhaps be preferable to an attempt to start over again. In fact, I have recently introduced in the United States Senate a resolution calling upon the President to make a request to the Secretary-General that the Continental Shelf Convention be reopened for the purpose of establishing a precise boundary. Mr. Alexander also favors, as did the Stratton Commission and other students of the problem, an intermediate-zone formula which would in effect extend the jurisdiction of the coastal nation and at the same time apply international rules and any agreed fees to exploitation within the zone.

Although such a formula is certainly preferable to more comprehensive claims of jurisdiction outward to the continental rise, I believe that on balance, our national interests will be best served by relatively narrow boundaries on the seabed to insure continued freedom of the seas above it. It is not realistic to expect that jurisdiction on the seabed will not impinge on the waters above, and our shrinking planet is already overburdened with burgeoning claims that seek to restrict freedom of the seas. It is probable, however, that a boundary drawn at the two hundred-meter depth measure of the Shelf Convention, without reference to distance from the shore or other factors, would not find sufficiently wide acceptance.

Senate Resolution 33 proposes that the limits of national jurisdiction be set at a depth of 550 meters or fifty nautical miles, whichever is greater, from the baselines used to measure the breadth of the territorial sea. The basis for the depth of 550 meters is that the edge of the continental shelf is not known to occur at any greater depth—the shelf as I understand the term and as it was understood by the framers of the Shelf Convention. Such a formula is in keeping with the spirit and the letter of the 1958 Geneva Convention on the Continental Shelf. It also offers a workable accommodation between those who advocate a narrower zone, and those favoring a much wider one. It is not looked upon with favor by the oil industry, but they seem to me very short-range in their perception of their own best interests.

Those who advocate a more extensive definition cite the exploitability clause of the Convention as a legitimation of as far-reaching a claim as technology may permit. A reading of the legislative history of the convention makes it clear that if the framers had in mind any view of the shelf as other than a topographical feature the two hundred meter criterion would be meaningless. That measure is significant only as it denotes the average depth at which the continental shelf ends and the continental slope begins. One can make a strong argument that the exploitability clause is meaningful only as a possible extension beyond that average depth but still within the limits of the topographic shelf, that is, out to the 550-meter maximum depth. The topographical definition of the shelf, despite efforts of oil company geologists to rewrite legislative history, is also consistent with United States domestic legislative action on the continental shelf question.

The alternative criterion, the measure of fifty miles from the shoreline, provides an adequate area of control and protection to those nations with narrow or nonexistent shelves. It also limits the temptation of underwater flags of convenience which would encourage some nations, taking maximum advantage of a broader definition of the shelf, to lease drilling and other rights at favorable terms and without burdensome requirements of safety and pollution control. If the seas are our common heritage, so too will be any despoliation of the ocean's ecology.

I hope that such a formula proves acceptable to those nations now seeking more extensive boundaries of jurisdiction both on the seabed and above it. Theirs is a dangerous game in a world already so congested that further restrictions of freedom on the high seas will be protested, and quite possibly contested by force. Surely there is some other way in which the legitimate interests of nations now making such claims can more reasonably be assured.

Finally, the 550-meter/50 miles proposal has the advantage of simplicity. It is a clear and measurable line of demarcation between national jurisdiction and that of whatever regime may be established for the deep-ocean beds. I believe that if this formula were adopted as a general rule, subject to modifications, problems such as that of islands would not be insuperable.

I hope that this meeting and those that follow will seek out the elements of accommodation in the many conflicting views of the future of the seabed. What is needed is a system that will encourage those nations and their nationals with the capacity to do so to exploit the seabed, to do so under agreed regulations for the preservation of our marine ecology, and to provide a means whereby some share of the proceeds may be of common benefit. Proposals that do not come to grips with political reality are likely to achieve none of these.

I would add one caution: we delude ourselves if we believe these problems can be solved in international negotiations before or in the absence of policy determinations within our own governments.

The time is short and getting shorter. The distinguished participants in this conference and in the meeting at Malta can offer much in pointing the way to acceptable compromise and in exemplifying the necessary ingredient of good faith. The alternative to effective international agreement will surely be a technological power scramble in which none of us, weak or strong, rich or poor, can be entirely certain of the outcome.

HAMILTON S. AMERASINGHE

16 The Third World and the Seabed

Developments in science and technology transcending even the most imaginative forecasts of fiction are making this one of the most exciting periods in the world's history. The statesmen of the world and the nations they represent are now called upon to exhibit the moral capacity that will be necessary to match the challenge presented by man's mastery of the physical domain. This is a great test of human statesmanship as well as an unrivaled opportunity for cooperative international effort.

The Third World and that new environment of human exploration, human activity, and human cooperation, the seabed, are most relevant to the task that confronts the statesmen of our day, and in this paper it is on these two factors that I wish to focus attention.

In international jargon the term "Third World" describes a group of countries which have chosen a deliberate policy of detachment from the cold war and avoidance of military alignment with either power bloc. Their philosophy has been described as nonalignment. It is a policy born of the conviction that peace and international security and the principles and purposes of the Charter of the United Nations cannot be attained by the division of the world into two groups mutually hostile and suspicious of each other; two groups impelled by that hostility and suspicion to intensify their preparations for war as the sole means of ensur-

HAMILTON S. AMERASINGHE is the Ambassador of Ceylon to the United Nations. He is Chairman of the United Nations Seabed Committee.

ing peace. Countries adopting this attitude or policy of nonalignment fear that every addition to the membership of one or the other of the two power blocs can only result in the intensification of international tensions. Between these two there had to be interposed a moral force built on the principles of equality, freedom, understanding, cooperation, and, above all, justice.

For some years now the United Nations has had under consideration the question of the reservation, exclusively for peaceful purposes, of the seabed and the ocean floor and the subsoil thereof underlying the high seas beyond the limits of national jurisdiction and the use of their resources in the interests of mankind.

The question was first entrusted by the General Assembly to an *Ad Hoc* Committee of thirty-five member States which was required to study it in its various aspects. The General Assembly considered the *Ad Hoc* Committee's Report at the 1968 session and decided to establish a Standing Committee of forty-two member states with a wider mandate than the *Ad Hoc* Committee. While the *Ad Hoc* Committee was required only to study the question in all its aspects, the Standing Committee was enjoined not merely to study the question further but also to make recommendations covering:

(a) legal principles and norms which could promote international cooperation in the exploration and use of the seabed and the ocean floor and the exploitation of its resources;
(b) ways and means of promoting the exploitation and use of the resources of this area for the benefit of mankind;
(c) the establishment in due time of appropriate international machinery for the promotion of exploration and exploitation of the resources of the area and the use of its resources in the interests of mankind irrespective of the geographical location of states and taking into special consideration the interests and needs of the developing countries;
(d) exploration and research in the area and international cooperation to that end and the stimulation of the exchange and dissemination of scientific knowledge on the subject;
(e) the prevention of marine pollution resulting from the exploration and exploitation of the resources of the area; and

(f) the reservation exclusively for peaceful purposes of the seabed and the ocean floor, taking into account the studies and international negotiations being undertaken in the field of disarmament.

As the title of the item and the content of the resolutions establishing the *Ad Hoc* Committee and the Standing Committee clearly indicate, the General Assembly's concern centered on the military possibilities and the economic potential of the seabed and the ocean floor beyond the limits of national jurisdiction.

In 1945 a Proclamation was made by President Truman which may be described as the first declaration of a claim of title to a part of the resources of the seabed and ocean floor. The Truman Proclamation asserted the principle of the national jurisdiction of coastal states over the resources of the continental shelf. It had one important direct consequence: international consideration of the limits of the continental shelf leading to the adoption of the 1958 Geneva Convention on the continental shelf. In the interval, and as a direct consequence of the Truman Proclamation, came the declaration by certain Latin American states extending their territorial waters to a limit of two hundred miles, a far call from the traditionally accepted limit of three miles. The two hundred mile limit of territorial waters was proclaimed by certain coastal states in Latin America which were, so to say, geologically disinherited in that they had no continental shelf and, therefore, considered themselves by way of compensation entitled to the living resources of the sea in as wide a maritime zone adjacent to their coasts as possible. The two hundred mile limit of the territorial sea carried with it a claim to the resources of the seabed underlying that extent of sea, a claim that is to a large degree of more academic interest than practical value. Whether claims to territorial waters extending beyond the traditional limit of three miles or to the currently widely-accepted limit of twelve miles are recognized or not, the existence of such claims has an important bearing on the problem of securing international agreement in regard to a clearer and a more precise definition of the continental shelf than is at present provided in Article 1 of the Geneva Convention.

In July, 1966 President Johnson made a memorable pronouncement which represented in advance the concern later expressed by the world community in regard to the use of this area and its resources. He stated on that occasion:

> Under no circumstances, we believe, must we ever allow the prospect of rich harvest and mineral wealth to create a new form of colonial competition among maritime nations. We must be careful to avoid a race to grab and to hold the land under the high seas. We must ensure that the deep seas and the ocean bottoms are, and remain, the legacy of all human beings.

This was an eloquent and inspiring expression of the principle that the fabulous resources of the seabed and the ocean floor should, instead of being the prey of international competition in furtherance of selfish national interests, be harvested and used for the benefit of all human beings.

The General Assembly item goes beyond the principle asserted in President Johnson's pronouncement and embodies the international community's anxiety lest world peace and security be further jeopardized by the competitive exploitation for military purposes of the strategic potential of the seabed and the ocean floor beyond the limits of national jurisdiction.

The Third World's philosophy is founded on principles which depend for their realization on the maintenance of conditions of peace and international security and recognizes the converse proposition that the application of these principles itself constitutes an essential prerequisite to international order. General and complete disarmament is the ultimate expression of man's faith in the peaceful professions and intentions of his neighbors and can result only from the elimination of the causes of tension and war, which are avarice, lust for power, mutual suspicion, and mistrust on the part of nations. In the search for agreement on disarmament the United Nations had endeavored, as a precautionary measure, to avoid the use of certain areas of the human environment for military purposes. The reservation of the seabed and the ocean floor exclusively for peaceful purposes was not, therefore, a novel idea.

The Antarctic Treaty of December, 1959 declared that Antarctica shall be used for peaceful purposes only and that any measures of a military nature such as the establishment of military bases and fortifications, the carrying out of military maneuvers and the testing of any type of weapons, as well as nuclear explosions, were prohibited in Antarctica. The same principle was later applied to outer space and under the Treaty on Principles Governing the Activities of States in the exploration of Outer Space, including the Moon and other Celestial Bodies, adopted by the General Assembly in December, 1966.

By comparison the military possibilities of Antarctica and outer space were more remote than those of the seabed and the ocean floor, especially since the use of outer space for military purposes seemed beyond the capacity of all but two of the major powers. On the other hand the seabed and ocean floor were a treasury of mineral resources which the progress of technology was making more and more accessible to the technologically advanced nations of the world. There was no real analogy between outer space and this new environment.

At the time the *Ad Hoc* Committee on the Seabed and the Ocean Floor began its study there was abundant evidence that the threat of extension of the arms race, and particularly of the nuclear arms race, beyond the limits of territorial waters into the area of the seabed and the ocean floor was no longer a matter of speculation but was assuming gravely disquieting proportions. It was evident that scientific research and development in regard to the seabed and the ocean floor had received its principal stimulus from the military possibilities of that area and from considerations of military strategy. It was those aspects of oceanographic research and development that were essentially relevant to and dictated by military exigencies which had attracted a growing volume of funds and an increasing commitment of manpower and technical resources. Most insidious of all was the danger of chemical and biological warfare carried out in the area of the seabed and ocean floor through the poisoning of plankton and the use of warheads carrying chemical and biological weapons whose lethal power would know no limits and respect no persons.

With these developments impending, the primary responsibility for arresting such a fatal trend and of preventing the extension of the arms race to the seabed and the ocean floor rests with the countries which have the technological and financial capacity to participate in the race, chiefly the Soviet Union and the United States. But their obsession with security and military considerations would probably impel them to advance research and development directed towards the nuclearization of the seabed and the ocean floor to the stage where agreement on the reservation of the area exclusively for peaceful purposes would be impossible of attainment. The Third World countries, in contrast, have an unqualified interest in the reservation of the area of the seabed and the ocean floor exclusively for peaceful purposes, and in the prohibition of all forms of military activity in the area which could possibly interfere with the exploitation of its resources. The political and military objectives of the policy of non-alignment pursued by the Third World coincide in this virgin zone where the problem is not one of disarmament but of the prevention of armament.

The preamble to the United Nations Draft Treaty on the Prohibition of the Emplacement of Nuclear Weapons and other Weapons of Mass Destruction on the Seabed and the Ocean Floor and the Subsoil Thereof recognizes the common interest of mankind in the progress of the exploration and use of the seabed and the ocean floor for peaceful purposes. This recognition falls short of what the Third World countries and many others desire, which is the reservation exclusively for peaceful purposes of the seabed and the ocean floor. That desire should be expressly stated and there should also be in the operative part of the Draft Treaty a categorical commitment that the efforts of the parties to the Treaty, in their future negotiations, will be directed towards a comprehensive prohibition of military uses in the area outside the maximum contiguous zone.

The reservation of the area of the seabed and the ocean floor exclusively for peaceful purposes, which has as its purpose the orderly exploitation of the resources of the area, would not be effectively ensured if military activities which would interfere with such orderly exploitation were permitted within the area. The proposal of the cochairman of the Conference of the Committee on Disarmament would exclude from the area only nuclear weapons or any other types of weapons of mass destruction as well as appurtenant structures, installations, and other facilities. To the argument that the prohibition does not eliminate military uses through the employment of other types of weapons, the answer has been given that only weapons of mass destruction could have sufficient military significance to warrant the expense of operation from the seabed and the ocean floor and that, therefore, any weapon that could have sufficient significance militarily to warrant the expense of operation from the seabed and the ocean floor would, by definition, have to be a weapon of mass destruction. Whatever force there may be in that argument, the types of weapons banned should be so defined as to exclude all possibility of the military use of the area outside the maximum contiguous zone. Also, weapons of mass destruction should be defined so as specifically to include chemical and biological weapons.

The proposal of the cochairman of the Conference of the Committee on Disarmament raises some doubt whether the armaments prohibited include submarines with nuclear capability or with the capacity for mass destruction. It might be argued that they fall within the category of facilities specifically designed for storing, testing, or using nuclear weapons or any other types of weapons of mass destruction, and that they are thus among the prohibited items. This would need clarification. Even the temporary use of the seabed and the ocean floor by submarines equipped with nuclear capability or with weapons of mass destruction could interfere with the peaceful use of the seabed and the ocean floor and should, therefore, be covered by the ban.

Those who desire the area to be reserved exclusively for peaceful purposes and wish military uses to be prohibited in the area cannot be satisfied with anything less than a total prohibition on the establishment, beyond the maximum contiguous zone, of any military base, fortification, or similar installation.

One of the brightest features of the Draft Treaty on demilitarization is the extent of the area in which the ban on nuclear weapons or weapons of mass destruction is to apply, namely, the area outside the maximum contiguous zone of twelve miles in width. This maximum contiguous zone may prove to be narrower than the area of national jurisdiction adopted for the purpose of exploitation of the resources of the continental shelf. This is an admission of the vital importance of limiting the area of the seabed and the ocean floor within which military activity may be permitted. We hope that the next step, which should be to secure agreement on a comprehensive ban on all military activity outside the maximum contiguous zone, will not present serious difficulty.

Implicit in the Third World's philosophy is the progressive reduction of the economic disparity between the developed and developing sections of the world. The United Nations as a whole has accepted the principle that the resources of the seabed and the ocean floor and the subsoil thereof beyond the limits of national jurisdiction should be exploited in the interests of mankind. The interests of mankind should be interpreted not merely as a vague ideal but as one of strict practical significance. It is not to be interpreted merely as the general interest and well being of humanity as a whole, which, it might be contended, could be served by the unrestricted and unregulated exploitation of the resources of the area just as any of the great discoveries or inventions may be said to have benefited all mankind. It should rather be seen as requiring a form of regulated exploitation of the resources of the seabed and ocean floor and the distribution of the income derived from such exploitation in a manner designed to reduce the economic disparity between the developed and developing sections of the world.

Despite all efforts, international and otherwise, the disparity in living standards between the developed and the developing countries continues to grow. If the vast resources of the seabed and the ocean floor were left open to the free and uncontrolled enterprise of nations possessing the technology and capital resources for their exploitation—and these would be the developed nations of the world—yet another source of wealth would have been thrown open to the developed nations and the plight of the developing countries would be rendered infinitely worse. The exploitation of deep-sea mineral resources otherwise than by international arrangement and without any understanding to the effect that the benefits would accrue chiefly to the developing countries of the world would have a disturbing impact on world trade and prices and on the economies of developing countries which depend for their foreign income on similar mineral resources and which would be faced with a new source of competition in world markets.

The principal mineral resources which are likely to be threatened by the products of the seabed and the ocean floor are manganese and phosphorites. Manganese nodules found in the seabed and the ocean floor contain varying concentrations of other metals such as cobalt, nickel, and copper. Apart from these minerals, petroleum and natural gas resources in the seabed and the ocean floor would also be available for exploitation. International arrangements must be made without delay to prevent the entry of the mineral resources of the seabed and the ocean floor into the world market in quantities and under circumstances which would have a depressing effect on the markets and economies of developing countries.

The first requisite of international regulation is the determination of the legal status of the area and its resources. The legal status must be such as to recognize the area and its resources as the property of the world community. The Third World would wish to secure international agreement that the area and its resources are regarded as the common heritage of mankind. This is the only status consistent with the objective of peaceful use and exploitation of the resources of the area in the interests of mankind. The concept of the common heritage of mankind is in effect the concept of property held in trust for all mankind, property which belongs to no single nation but to the entire world. Although it concedes universal ownership, and therefore fair and equal shares for all, considerations of equity demand that priority in the allocation of the proceeds of exploitation should go to the neediest in the world and that would be the developing nations of the world.

The administration of this trust on behalf of all mankind would call for the establishment of some form of international organization and machinery which would manage the area, regulate the exploitation of its resources, the marketing of its products, and the disbursement in an equitable manner of the income so derived.

Various forms of management of the area and its resources have been suggested. Those who are averse to any form of international control because it raises the specter of a supranational administration would want a simple system of registration of claims or interests by individual nations or even enterprises. Under this system the contemplated international authority would function merely as a central registry. This would scarcely be acceptable to the Third World. The seabed and the ocean floor, so far a sealed treasure chest, would under such a system of free registry become a Pandora's Box, two extremes which must be avoided. It would also be inconsistent with the concept that the seabed and the ocean floor and its resources beyond the limits of national jurisdiction are the common heritage of mankind, and with the objectives stated in the General Assembly Resolutions and fundamental to the Third World policy, of the progressive reduction of economic disparities between the developed and the developing sections of the world.

The objective of peaceful exploitation in the interests of mankind with special regard to the needs and interests of developing countries of the world could best be attained by the creation of an international authority which, through a system of licenses or contracts, or both, together with the financial concomitant of royalties and fees, would provide for the orderly exploitation of the resources of the area and for the application of the proceeds to the economic development of developing countries and to other international community purposes. The constitution providing for the establishment of such an international authority would have to incorporate safeguards against the domination of the organization by any special interests or group of interests, or by any particular country or group of countries, and would also have to create conditions under which a system of licenses and contracts would be ensured free from all discrimination. The international agency would have to be empowered to take appropriate measures to ensure that the marketing of the products of the area would not prove detrimental to the interests of developing nations and to adopt compensatory financing programs or policies for the relief of nations deprived of a part of their traditional markets and of their foreign exchange earnings through the operations of the international agency itself. Underlying these proposals is the fundamental proposition that the sole right of ownership will remain vested in the world community and be inalienable.

The issues involved are highly controversial and contentious. International agreement on the means of regulating and controlling the activities of the seabed and the ocean floor and the exploitation of its resources in the interests of mankind through an international agency is not likely to be reached without long and strenuous negotiations. In the meantime, it is of the highest importance and in the interests of international peace and security that the nations of the world should with the least delay adopt a set of principles recognizing the proposed legal status of the area and its resources as the common heritage of mankind, and asserting the principle that no state may claim or exercise sovereign rights over any part of the area and its resources and that no part of the area and its resources shall be subject to national appropriation by claim of sovereignty by use, occupation, or by any other means. The General Assembly moratorium on the assertion of such claims and on the conduct of certain activities in the area which would be prejudicial to the realization of these principles and the objectives embodied in the General Assembly Resolutions on the subject is a significant step.

However, any agreement on principles must be accompanied by clear definition of the area of the seabed and the ocean floor over which a state may exercise national jurisdiction. The only definition existing today is contained in the 1958 Geneva Convention on the Continental Shelf, which suffers, however, from the grave defect of ambiguity. It is in the interest of the Third World to secure revision of that definition. A precise definition of the continental shelf would enable the limits of national jurisdiction to be clearly ascertained.

The international authority, the establishment of which is being proposed, would need to be vested with responsibility for preventing the pollution of the area, for ensuring freedom of scientific investigation and research without discrimination or without any claim to exclusive use arising from any research activity, for promoting international cooperation in all such scientific investigations and research, especially with the object of disseminating among all states the results of such research and investigation, and for providing technical assistance and facilities to developing countries for the training of personnel.

For the present purposes it is not necessary to consider in detail the means by which the resources of the area could most efficiently be exploited. It is necessary to identify general principles and objectives compatible with and conducive to the realization of the policies and principles underlying the Third World's philosophy. The relevant resolutions of the General Assembly give a clear indication of the objectives in contemplation. A unique opportunity exists for international cooperation in a manner and on a scale calculated to ensure the peaceful exploitation of the resources of this new environment, and orderly and even economic growth throughout the world. It would be a gigantic experiment in international cooperation and an endeavor that could herald a new era of growth for the developing nations so long condemned to tragically slow progress dependent on the bounty of the rich. Recognition of the wealth of the seabed as the birthright of man, and its use to raise the standards of living among the less developed countries may well prove to be this century's most significant achievement in the cause of peace and a grand overture to the new millenium.

ARVID PARDO

17 New Horizons in Ocean Science and Law

Science and technology are producing the most significant revolution in the history of mankind. A revolution in our perspectives, a revolution in our civilization, a revolution in our concept of power. We are on the threshold of a new age of discovery. The dimensions of our world are exploding. On the one hand, we are reaching to the planets, on the other, to the mysterious depth of the oceans. This is paralleled by a revolution in our expectation from, and in the dimensions of, our life. Advances in medical science, biology, and physics suggest that in a not too remote future it may be possible greatly to prolong useful human life and to create new forms of life, perhaps even a new man. Recent progress in controlled thermonuclear reaction, superconductivity, magneto-hydrodynamics, communications, engineering, and a wide range of scientific fields, some totally new, are creating a new civilization that is integrating human activities into a global geo-technical whole which within this decade will encompass both the oceans and the underdeveloped regions of the world. Finally, science and technology are giving us undreamed-of power in the military but even more in the civilian field. This process is immensely broadening the vistas of mankind. At the same time it is increasing immensely the dangers and hence the responsibilities which must be faced by mankind. Science has no moral con-

---

ARVID PARDO is Malta's Minister Plenipotentiary for Ocean Affairs at the United Nations. He originated the proposal in the United Nations for an international seabed regime in the United Nations.

notation and technology is an unpredictable variable. It produces interdependence of nations and regions but does not necessarily produce unity. On the contrary, it can produce (indeed it has in the past) conflict both within and between nations and regions. The discovery of anti-matter has given us new precious insight into the cosmos. Within a generation, however, it may give some nations the awesome power to destroy not only life but our very planet. The breaking of the genetic code and the infant science of genetic engineering offer unfathomable opportunities for improvement in the quality of human life. They may also someday place in the hands of governments the power to create a race forever destined to slavery.

Our present rudimentary capability for weather control has the potential eventually vastly to improve agricultural productivity. Weather control technology, however, could also assist us in acquiring the capability to impair our atmosphere, for instance, by interfering selectively with the ozone layer surrounding the earth, thus exposing the ground below to the disastrous effects of direct ultraviolet solar radiation.

Our intrusion into the ocean deeps will enable us to supply more abundantly the needs of growing industries and multiplying populations. It also gives us the power irreparably to contaminate an environment essential to life. In sum: technology can unite and it can divide. It can elevate and it can degrade. It can create a new civilization of abundance, it may destroy all civilization and life on this globe. The speed of technological innovation is accelerating. Its scale and cost is increasing in geometrical proportions, imposing a heavy burden on even the richest countries. This imposes upon us not only an unprecedented effort of adaptation and imagination but also threatens to make of that part of the world that cannot sustain the present pace of technological change a submissive object of manipulation. Thus by the end of the present decade, we will face the most momentous crisis ever experienced by mankind in its million years of history. At stake is the survival of man himself. The problems basically are how to control technology and its impact and *who* will control technology and its impact. This raises the problem of institutions both at the national and international level. The inability of many of our existing institutions to make efficient use of science and technology and to channel them into constructive purposes is perhaps one of the basic causes of contemporary unrest. Since the dimensions of technological impact range from local to global, control, in the last analysis, can be effective only if undertaken on a world-wide scale. Yet the political and institutional organization of the contemporary world does not permit such control at the present time. This is the tragedy; this is the challenge to all of us.

It is against this general background that my country submitted its proposals with regard to the seabed nearly three years ago. These were a small but essential part of a comprehensive concept. They should be viewed as an attempt by the second-smallest country in the United Nations to make a modest contribution to the world order which must emerge if we are to live in peace and dignity, indeed, if we are to live at all.

The suggestions made by the Government of Malta were extensively debated in the United Nations and the debates have been extremely useful. Progress has undoubtedly been made with the assistance of the many useful studies prepared by the Secretariat of the United Nations, particularly in the understanding by governments of the complex issues which were raised. A forum has been created where the question of the future status of the seabed can be discussed in a context which is political in the etymological sense, the deepest sense. An impulse to international cooperation in the exploration of the marine environment has been given by the United States initiative in the Seabed Committee to launch an international decade of ocean exploration: an initiative unanimously endorsed by the United Nations General Assembly. Negotiations have been initiated and have made good progress with regard to the question of the prohibition of the emplacement of nuclear weapons and weapons of mass destruction on the seabed. There has been agreement also on some general points although the substance of these points still remains quite controversial. Thus consensus now exists that an area of the seabed beyond national jurisdiction does in fact exist but there is disagreement as to its limits. There is general agreement that the seabed beyond the limits to be agreed upon should be reserved for peaceful purposes but there is disagreement on the meaning of the words "peaceful purposes." There is agreement that pollution of the marine environment must be controlled and if possible avoided. But there is disagreement on the methods of control and on whether states should be held liable for serious damage caused by them to the marine environment. There is agreement that the seabed should be exploited for the benefit of mankind as a whole but there is strong disagreement on the practical content to be given to this phrase. There is a very wide feeling that some type of international institution should be established but there is equally wide disagreement on the functions and powers of such an institution.

Lack of agreement on three difficult, complicated complexes of problems appear to be the main obstacle to progress.

The first problem, or complex of problems, concerns the nature of the basic concept that should govern the exploration, use, and exploitation of the seabed beyond national jurisdiction. Should it be an adaptation of the traditional concept of the high seas, or should it be the new concept of the common heritage of mankind? The second difficulty is the nature of the regime and of the international institutions which it is proposed to establish. Can an appropriate, equitable, and effective legal regime for the seabed beyond national jurisdiction be established without creating international institutions? If international institutions are necessary, would their competence extend to the management of the seabed beyond national jurisdiction including regulation of all uses and supervision perhaps of military uses, or should the competence of international institutions be limited to the question of resource exploitation or perhaps even only to a mere registration of claims to exclusive exploitation rights over certain areas? In the event that it were found desirable to create international institutions with a wide competence and strong powers, how can the conflicting interests of States be balanced in such a way as to insure both the viability of the Regime, which must include the respect of the vital interests of all states together with the satisfaction of the needs of the poorer countries, and the impartiality and efficiency of whatever international institutions are created?

Finally: what are the limits of national jurisdiction, or to put the matter in another way: what are the limits of the area beyond national jurisdiction? The international debate has scarcely started on this question which involves a difficult revision of some of the articles of the 1958 Geneva Convention on the Continental Shelf.

The three problems which I have mentioned are closely interconnected and involve issues of quite fundamental importance to most states. It is not surprising therefore that debates in the United Nations have been prolonged, that all aspects of every question have been minutely examined, and that action has been slow. States are aware that, in the words of Eugene Skolnikov, "the precedents established now will be critical elements not just to determine a regime for the oceans for the next few years but in laying the basis for the future organization of an increasingly interdependent world." This is an immense, most difficult and complex task which will test all our imagination, all our idealism, all our realism. But it is also a most urgent task. The pace of technology will not wait. When the appropriate technology is available, it will be employed, since this is the logic of international competition.

I am most deeply grateful to the Center for the Study of Democratic Institutions for holding in Malta this meeting which brings together experts from so many countries and from so many different disciplines. I am hopeful that this cross-fertilization will produce new insights and that these will assist the world community to give itself the international institutions which are, and increasingly will be, required by the progress of science, the impact of technology, and the pressing needs of man.

PART FOUR:

PACEM IN MARIBUS

The provisions an international ocean regime must make to insure Peace in the Oceans—Pacem in Maribus, are of two kinds: It must provide for disarmament and arms control and it must provide inspection, constabularies, and enforcement mechanisms concerned with the nonmilitary uses of the seas.

Thus far, these two aspects have been painstakingly and painfully kept apart. The CCD dealt with disarmament; the Seabed Committee, with peaceful uses. Consequently the progress of the CCD negotiations has been slow, and the result disappointing, to put it mildly. Elizabeth Young's and Alva Myrdal's analyses of the Treaty—which, in the meantime, has been adopted by the General Assembly—certainly bear out the criticism offered in the introduction to this volume. In view of the military hardware, available now or in the foreseeable future, as described in the chapters both of Elizabeth Young and Neville Brown, these shortcomings of the Treaty are fraught with catastrophe.

Elizabeth Young posits as one of the conditions for successful arms control that the "control should be part of ordinary ongoing processes useful for other purposes as well."

Frances and Robin Murray point out that constabularies concerned with the nonmilitary uses of the sea should be taken into due account in the course of developing verification measures for marine arms control. Alva Myrdal suggests that if and when an international machinery for the seabed were set up, it might be possible for states so desiring to make use of that machinery for the verification needs in relation to the disarmament Treaty.

It is possible that it is this very connection—and only this connection—between disarmament on the one hand and the development of peaceful uses of the oceans on the other that distinguish the ocean case from any other in the past and that make disarmament possible. This is what is meant by a peace system.

Both functions—arms control and the policing of peaceful uses —might eventually be entrusted to an "ocean guard" as proposed by Senator Pell in his Senate Resolution 33. Whether this should be created ex novo *or whether it should use and coordinate existing control forces, international and national, public and private, as they are described in the Murrays' chapter, is another question. Quite likely, this latter alternative may turn out to be more economical and more realistic.*

Although military integration touches on the nerve center of national sovereignty, there are numerous precedents for some degree of military integration such as the Warsaw or the Atlantic Treaty Organizations. In the past such integrations, with or without political union, have only occurred when communities wanted to defend themselves against a common threat from outside. *If this were necessarily so, the world community could never unify either its military controls or its political structures, since it has no outside against which to defend itself in common.*

Curiously enough, things seem to be working out in a different way in the world at large: it is not to defend ourselves against the outside that we move toward unification; we do so to defend the outside (the environment) against ourselves. To defend the environment from pollution, nonmilitary or military, requires coordinated action on a universal scale.

The internationalization of controls in a peace-system offers the only hope to developing nations for the safety and health of their own coastal zones. Ocean policing is an expensive affair requiring a highly developed technology. The wider the claims of the developing nations for national sovereignty out in the oceans and on the seabed, the more difficult and the more costly it would be for them to control safety and antipollution standards in their coastal zone—for which they would be responsible under international law. The greater the area over which they have sovereignty, the greater their dependence on the international community. This is one of the paradoxes of the situation.

The following chapters review the military uses of the oceans, the situations where conflict may arise, the existing political instruments to cope with it, and the real needs of an international Ocean Regime. National militarization and international peaceful exploitation of the riches of the seabed simply cannot coexist.

LORD RITCHIE-CALDER

18 In Quiet Enjoyment

Seven-tenths of the surface of our planet are covered by the waters we call oceans. The other three-tenths are the continents which, so far, have provided the living-space and the material needs of mankind. On this land-mass, *Homo sapiens* evolved, migrated and settled. The species differentiated into ethnic groups that carved out territories which became nation-states and which were subdivided into properties, landed estates, homesteads, fields, and urban realty.

The rocks and subterranean strata of the continents were quarried, mined and drilled to extract the solid minerals and the liquid oil to fabricate the material needs and provide motive power. Those resources also represented property to be claimed, exploited, and protected against counter-claimants and expropriators. To secure the "quiet enjoyment" (to use the lawyers' phrase) of such properties, an elaborate system of laws had to be established and had to be supported by constabularies, by national armies, and, in our day, by the massive, long-range, armaments of global strategy.

The oceans, all 140,000,000 square miles of them, were the waters which separated the continents and the islands, which provided the thoroughfares for trading ships and, in war, the battle-

---

LORD RITCHIE-CALDER of Balmashannar is a Senior Fellow of the Center for the Study of Democratic Institutions. He has worked as a journalist, with the United Nations, and as a professional science writer. Formerly he was Professor of International Relations at the University of Edinburgh and Chairman of the Metrication Board of the British Government.

grounds for navies. Periodically, nations, with a sense of naval supremacy, such as the Portuguese, the Spaniards, or the English, would claim dominion over areas of the open, or high seas. (The Spanish Main was the South and Central American mainland bordering on the Caribbean Sea or vice versa.)The English jurist, John Selden, in 1635, expounded the legal doctrine of *Mare clausum* in which he asserted that "the sea by the law of nature or nations is not common to all men but capable of private dominion or property as well as the land." This, however, did not prevail against *Mare liberum* in which Hugo Grotius, the Dutch jurist, in 1609, embodied the doctrine of Freedom of the Seas, qualified only by the practical need of a coastal state to exercise some jurisdiction in the waters adjacent to its shore. In the eighteenth century, this was defined by Van Bynkershoek as the actual distance which could be protected by land-based cannon. This range (overambitiously for the weaponry of the time) was defined as three miles.

Beneath the surface of the seas, on which ships had the right of free passage, there were fish. As fishing boats extended their ocean-going capacity and could reach fishing-grounds farther and farther from their own coasts, coastal nations sought to safeguard the livelihoods of their own fishermen by protecting their rights within territorial waters and by seeking to extend the limits of those waters. This took the form of armed protection against foreign competitors as well as a marine constabulary to ensure good fishing practices by their own nationals. Furthermore, the navies of the maritime nations, in common interest, provided a form of collective security against piracy on the high seas.

Naval activity, aggressive or defensive, entered a new dimension with the advent of the submarine, which could operate in the concealment of the covering waters.

With the increased efficiency of aerial and satellite surveillance systems—the extension of the aerial photography which had revealed in detail the construction of the Cuban missile sites—and with the target accuracy of nuclear weaponry, fixed land bases had become vulnerable. The answer sought was mobility or concealment. The opaque depths of the sea offered both.

This new dimension of strategy intensified the research and

technology for operating at great depths. When military services "pick up the tab" and, through their appropriations, take care of research and development costs, the time-scale of technological innovation becomes radically different. The "spin-off" from an expense-no-object military program can become the "know-how" of civilian operations. The missile program became the satellite program; the bleeps of Sputnik I became the telecast of Man-on-the-Moon in 1969, twelve years later. With the cut-back of the outer space program in 1970, the military-industrial complex of the United States were looking for diversification and for employment of their "know-how" and their manufacturing capacity for purposes of inner space, the ocean bed.

Aerospace can become hydrospace in seeking alternatives. This is important in any discussion of the need of an Ocean Bed Regime because there has been a tendency to say "What's the hurry? Deep ocean technology has a long way to go." But has it? It is now quite clear that depths and pressures are no longer regarded as ultimate deterrents. Materials-technology is already far enough advanced to promise manned vehicles even at the greatest depths—although Man-in-Depth will only be incidental to most of the operations involved in the extraction of the resources of the ocean floor, or its subsoil. Dr. John Craven (at the Center for the Study of Democratic Institutions seminar in January, 1970) said, with the assurance of unique experience: "It is technically feasible to put men, material and equipment in the deepest part of the ocean and it will be feasible at low cost in the very near future."

If, therefore, this "know-how" is applied either for military purposes or by some adventurous enterprise which badly wants something of resource-value from the sea-bottom, part of the common heritage of mankind will be preempted; not only will we be wrestling with "squatter's rights" but, in the absence of a regime to require proper safeguards, exploitation may cause serious, perhaps irreversible, ecological damage.

The agencies of national governments can supervise and regulate possible hazardous practices on the Continental Shelf but there is no present body which can lay down rules of behavior and procedure for those who on their own initiative might seek to exploit the deep sea bottom.

While the discussions in the United Nations, now transferred from the *Ad Hoc* Committee on the Seabed to a Standing Committee continued, the U.S.A. and the U.S.S.R. in October 1969 agreed to a Joint Draft Treaty within the context of demilitarization. In this they declared themselves willing not to emplant or emplace any object with nuclear weapons or any other type of weapons of mass-destruction in or on the ocean floor. It seemed such a nice gesture but it was received with some ungraciousness by many nations. It simply meant that the strategists of the two major powers had already discarded the idea of ocean bottom fortresses in favor of evasive mobility. Strategical submarines of the Polaris type or other forms of submersibles had been accepted as a better proposition.

This underlines the obvious, that the neutralization of the ocean bed, its floor or its subsoil, cannot be considered apart from the suprajacent waters or from the offensive-defensive uses of the Continental Shelf. The exercise of treating it as a special aspect, however, was invaluable because of the novel issue which it raises and the new frame of reference it provides for security and arms control.

That was why the Center for the Study of Democratic Institutions in preparing the materials for the Malta conference, Pacem in Maribus, made the symposium on demilitarization of the seabed its first in the series. It served both its own and the general purpose, because it raised pretty well all the issues with which one has to deal in considering a possible Ocean Regime—even to the point of making the need for such a Regime not "possible" but imperative, and urgent.

The examination, with experts available, showed, as has already been stressed, that the time-scale in which development of resources for peaceful purposes and, yes or no, the common good of mankind, is considerably modified by military incentives. If there is a military impetus behind the technology, access to material and nonmaterial resources may not be deferred as long as some present estimates would suggest. Strategic minerals in short supply in a nation's inventory could themselves provide a pretext for unilateral exploitation. That would raise the question of how right of access could be acquired, or sustained, in the absence of some regulatory body. It also raises questions of possible damage to the ocean ecology if indiscriminate development should take place, whatever the pretext.

At one stage, our discussions in the symposium seemed almost like a script conference for a wild western scenario. We were opening up "The Last Frontier," the oceans, with adventurers going out into the virgin territories staking their claims and repelling interlopers, until the federal marshal came along to represent law and order, followed by the elected sheriff and the appointed constabulary of a regime of law-and-order. The wild western analogy could, comfortably, be extended to the conflict of other interests—the cattlemen driving out the hunters, the sheepmen in conflict with the ranchers, and both resenting the homesteaders and the competing claims of the railroad "barons" and the primacy of oil. This can be correlated with the conflict of interests between the extraction of material resources and the fishing of living resources and with the freedom of marine transport. When we tempered that melodrama, we still had symptoms worth examining. We looked at present tendencies to see the future challenges. The giant tankers are a case in point: they are not only an ecological hazard but an insurance risk which the tanker companies have now corporately to underwrite. This can only mean an inspectorate to ensure that the conditions are observed and, in the case of pollution, some sort of self-policing system to see that waste oil is not irresponsibly dumped. There is already the example of TOVALOP, The Tanker Owners Voluntary Agreement Concerning Liability for Oil Pollution. This is a combination of private interests and from it is bound to emerge a new form of supervision and enforcement.

The role likely to be played by large corporations whether nationally incorporated or multinational, in relation to the sea "beyond national jurisdiction," raises issues which do not arise on the Continental Shelf where national jurisdiction applies. Through whom and with whom would they negotiate in the absence of an Ocean Regime? Or, if an Ocean Regime existed? How would their "quiet enjoyment" of rights be allocated and, having been allocated, how would they be ensured? It raises new questions of collective insurance and of the inspectorate and constabulary necessary to protect the "common heritage" from expropriation or pollution and how to protect rights once granted in relation to other rights, i.e., fishing, possibly fish farming, transport, and communications.

As will be seen from the chapters which follow, an "exercise" which started from the question of demilitarization of the ocean bed, led by an inescapable compulsion to an examination not only of present strategies, of problems of general disarmament and of the spin-off of military technology, but to the questions of a new kind of collective security, of a new custodianship, of a new constabulary, and of a new legal regime beyond the definitions of present international law. The fact that most of these were beyond the self-evident types of reference only illustrates the complications, the implications and the opportunities involved in the recognition that seven-tenths of our world is a "property" which has to be defined, developed, defended, policed, inspected and allocated. And if we add, "in the common interest of mankind," this requires some kind of active trusteeship which at the moment does not exist but must be contrived.

ELIZABETH YOUNG

19 Arms Control and Disarmament in the Ocean

In order to find a context for arms control on the seabed and in the seas beyond the limits of national jurisdiction we shall begin with a brief survey of existing and predictable technologies and a sampling of conflict areas. It will be useful to examine the existing experience of arms control in other areas and elements—what has and what has not been achieved, and also what has been discussed. The progress of the overall strategic balance also has to be related to this general arms control experience, and so do the ways in which sea and seabed arms control differ from the rest.

TECHNOLOGIES

(NOTE: *Almost all the references in this section are to Western sources. Soviet information on these matters is limited.* EY)

In ocean technology a large number of theoretical and production breakthroughs are more or less imminent: the fuel cell or some other small, autonomous, and longlasting source of power for small vehicles, including submersibles, is likely to be an eco-

---

ELIZABETH YOUNG graduated from Somerville College, Oxford (where she was an Exhibitioner) in Philosophy, Politics and Economics. She achieved the rank of leading Wren in the Women's Royal Naval Service. She has written widely on disarmament, and arms control, particularly in *The Bulletin of the Atomic Scientists* and in British papers. She is married to Lord Kennet, a junior minister in the present British Government with special responsibilities in environmental affairs. They have six children.

nomic proposition within the next five or ten years. Basically low cost materials for the construction of submersible vehicles, such as glass, kinds of fiber glass, syntactic foam (hollow glass miscropheres in an epoxy resin matrix) are already available. So, in theory, is free-flooded machinery. The U.S. Navy Deep Submergence Vehicles are themselves designed to be transportable by air, by the Lockheed C141, or on the "back" of an ordinary submarine.[1] A system of navigation satellite stations over each of the oceans, which first allowed missile-launching submarines to pinpoint their location, is now to be available to ordinary shipping too; and, for ship traffic control, including the precise location of shipping by the same satellite, a universal system could be in service by 1975.[2] Weather forecasting is about to become a science, because at last enough information will be available. Surveys by satellite of the earth's resources will also help the fishing industry by allowing schools of fish to be located and followed.[3] Enormous increases in the size and capability and speed of both conventional and unconventional, submarine and surface craft are on the way, including factory ships of various kinds for the fishing and oil industries.[4] The container revolution and the fact that a considerable proportion of the world's tramp fleet built during the second World War is due to be scrapped, are combining to produce an entirely new style in general ship building and management.[5] New methods are being developed for extracting oil offshore—some of them useable not only far out at sea but at ever increasing depths. The oil can be stored in huge, bottomless tanks into which the oil rises, and, as from a samovar, can be poured directly into tankers or bladders, quite independently of the shore.[6] Submarine tankers are proposed and so presumably are submarine terminals.[7] Techniques for fish and shellfish farming and rearing are progressing, for use both close to shore and at sea; dredging, trawling and mining for minerals on the seabed may for a time be slowed down by the satellite-aided discovery of further mineral resources on land, but the technology for obtaining them from the seabed is growing, either by drilling or trawling.[8] Certainly within the seventies, the construction of platforms and artificial islands will be a common economic proposition, whether on the surface or submerged, for airfields, nuclear and other power stations, and any other noisy, hot, smelly or unpleasant purpose.[9]

Some of these civil developments are in harness with the course of military developments at sea. A cheap fuel cell (or the equivalent) and cheap construction materials and techniques will allow the world's navies progressively to disappear into the vast spaces beneath the surface of the sea (cheap, that is, in comparison with nuclear propulsion). These non-nuclear techniques will probably provide the same kind of country as could achieve a small nuclear weapon capability, with a twenty-day submersible. Advanced navies will certainly go on developing the nuclear-powered submarine, the cost of which Mr. John Craven, for one, believes may come down to about half that now current. The Non-Proliferation Treaty, of course, does not control the transfer of fissile material for military purposes (as opposed to explosive devices—see Article III); and enriched fuel for nuclear submarines is likely to be available from several sources.

Military satellite observation systems and improved guidance systems for missiles will hasten the process by making surface ships, particularly large ones like aircraft carriers, increasingly vulnerable. So will the use of lasers against ship-based air defenses: [10] a laser beam will penetrate to a depth of a thousand feet, but not much further. The United States, and perhaps the Soviet Union, are carrying on research into a long-range underwater launched missile system (ULMS). Dr. Craven's view is that a design could now, though at very great expense, be built to fire from twenty thousand feet below the sea. Stable platforms will be used by the military for airfields, docks, both surface and underwater storage, and other traditional naval base functions.[11] They could also be used to carry antiballistic missiles and their related radars, for retrieval facilities for various kinds of information systems,[12] and so on. Sea floor engineering will be costly rather than difficult.

On the production side, in the United States the whole process, military and civil, will be pushed by the Administration's decision to limit the space and missile programs. This has resulted in substantial unused capactity in the great aerospace firms, and it is clear that many of them are now making equipment for use in the ocean. Several firms are producing prototypes of submersibles; and opinion on the world's stock exchanges seems to agree with them that navies and industry are indeed about to go submarine, and that this is likely to be a profitable field for investment. Something of the same sort may be going on in the Soviet Union, where the space program also appears to be in eclipse.

DISPUTES AND CONFLICTS

Many thousands of miles of uncertain new frontiers were created by the 1958 Convention on the Continental Shelf. This is already resulting in trouble, for signatories and nonsignatories alike, some of it threatening to explode in violence. One of the few "laws" established by political scientists is that conflict most easily erupts over frontier disputes.

Current disputes range from polite disagreements, as between West Germany and Denmark and the Netherlands over the drawing of boundaries on the bed of the North Sea,[13] to the threats the Egyptian government issued in November, 1969, over prospective Israeli-sponsored drilling for oil on the continental shelf some thirty miles offshore in the Gulf of Suez by an American company, Midbar, registered in London and using Canadian drilling equipment. Some reports suggest that the five gunboats that unexpectedly left Cherbourg last Christmas for Israel were intended "to protect offshore oil-finding operations."[14]

New frontiers will have to be drawn. That the United States and the Soviet Union now have a common submarine frontier should not lead to trouble. The division of the North Sea has proceeded peaceably enough. However, if there is oil in the Mediterranean, submarine boundary drawing will be very difficult indeed. Already there is disagreement between the United States and Canada about Canadian claims, not only to the islands but also to the waters of the Arctic Archipelago, as well as over mineral rights on the Continental Shelf between Nova Scotia and Maine. In Southeast Asia and the Far East, where oil has already been found, division of the seas will be an even worse proposition because of the great number of islands and states involved. The "lease map" of this area is now nearly full,[15] and within the next few years exploration will have found out just how valuable this particular bit of Continental Shelf is. The situation will be exacerbated by the fact that many governments hereabout do not recognize each other. One map now in circulation shows an area on what appears to be the North Korean shelf leased by the Government of South Korea to Messrs. Caltex for exploration.[16] Argentinian claims to the Falkland Islands are perhaps partly affected by the great expanse of their Continental Shelf; and, the British Government is currently examining the possibility of oil there. In the northern part of the Persian Gulf, disputed boundaries on land result in disputed boundaries on the Shelf. Iranian naval craft are said on at least one occasion to have forced an American company to move its gear from a disputed area. The Gulf of Suez dispute mentioned above is further complicated by the existence in the area of an oil field jointly operated by the Egyptian Government and Standard Oil of Indiana.

Disputes arise also about the nature and habits of various kinds of fish: the Brazilians have successfully maintained against the French that a particular kind of lobster is sedentary and may therefore not be taken by foreigners. Equally, the Russians are successfully maintaining the king crab to be sedentary, while the Japanese maintain it swims: the crab is not freely available to Japanese fishermen. Britons and Danes are in dispute at the time of writing because the salmon is at some stages of its life a river fish, and at others a sea fish: the British believe the Danes are over-fishing British salmon in the North Atlantic. There are other fishery disputes between Russians and Japanese, some of them consequent upon Russian occupation since 1945 of certain previously Japanese islands. The protection of traditional fishing rights, particularly in the face of the highly developed and competitive techniques of some fishing fleets, has caused several Latin American states to declare, and enact into their domestic law, territorial waters extending two hundred miles. The United States and Soviet Union are now jointly proposing a universal maximum of twelve miles.[17] Trouble is likely to ensue if any attempt is made to enforce such a twelve mile limit; for instance, by providing naval protection for fleets fishing within the Latin American limits.

If the owners of jumbo-tankers solve their insurance problems, disputes may arise over the desirability of "improving" certain international narrows. Thus, the Japanese government wishes to make a full-scale survey of the Straits of Malacca, through which more than ninety percent of Japan's oil now passes, much of it in tankers for which the Straits are dangerously shallow. The first reaction of the Malaysian government was to extend its territorial waters from three to twelve miles, thus acquiring a right to block the Japanese proposal. A compromise has now been reached, in which, no doubt, Malaysian interests will be protected.

If a general agreement were reached to universalize territorial waters at twelve miles, a number of international narrows would be affected, at least as far as the seabed is concerned. In some cases, advance permission might be required for the passage of warships.[18] Those which would then be entirely through territorial waters include, among others, the Dardanelles, the Straits of Malacca, the sound between Sweden and Denmark, the Straits of Dover, the Strait of Bab el Mandeb, the Strait of Gibraltar, the narrows at the entrance to the Persian Gulf and the Bering Strait.

EXISTING EXPERIENCE

Although disarmament and the regulation of armaments have always been a central concern of the United Nations, little has been achieved under its immediate auspices. However, the Partial Test Ban, the Space Treaty, the Non-Proliferation Treaty, and the Strategic Arms Limitation Talks, in spite of being negotiated elsewhere, have all sprung from requirements and alarms first effectively voiced in the United Nations.[19] Mr. Krishna Menon's speech of July 12, 1956 is the richest single vein of insight into disarmament issues of the whole nuclear postwar. In it he reiterated Mr. Nehru's proposal for a nuclear test ban. He referred to "what used to be 'the third country' problem. Now it is the 'fourth country' problem. Next time we meet it may be the 'fifth country' problem." He spoke of the importance of an actual first step in nuclear disarmament. He asked for no trade in nuclear weapons, a freeze on their manufacture, and no transfer of either nuclear weapons or fissile material. He suggested direct negotiations between the United States and the Soviet Union and that the Disarmament Commission should be more representative. He said, "control is the [expression] of the determination of nations for the securement of agreements that have been reached."[20] We are only now catching up with this.

SUBMARINE AND OTHER SHIP-LAUNCHED MISSILES[21]

United States

		In Service	*Range*	*Warhead*
Ballistic Missiles Submarine launched from underwater	Polaris I	1960	1380 st.m.	0.7 megaton
	Polaris II	1963	1700 st.m.	0.7 megaton
	Polaris III	1964	2850 st.m.	0.7 megaton
	Poseidon	1970		
Ballistic Missiles Surface launched from submarine	–	–	–	–
Ballistic Missiles Ship-launched (destroyers)	–	–	–	–
Cruise Missile, Ship-launched. (Cruisers)	–	–	–	–

The Secretary of Defense of the United States, Mr. Laird, in his report to the Senate Armed Services and Appropriations Committee on the Fiscal Year 1970 Defense Programme and Budget, mentioned that the United States was not "now" committing itself to (among other things) a Mobile Minuteman afloat.[22] However, research and development, at a cost of ten million dollars, is continuing for an Underwater Launched Missile System, which would be a successor to the Polaris/Poseidon System. For the Fiscal Year 1971, forty-four million dollars for research and development is being asked.[23] Comparable information about the Soviet Union is not available.

There are two unavoidable conclusions from an examination of the world's experience of arms control in the nuclear age: what has been achieved has been ineffective and insufficient, and it has been bipolarist in the sense that it has been devised and carried

Soviet Union

	In Service	*Range*	*Warhead*
Sawfly Missile A Polaris-type (in Y class sub.)	1969	1200-1500.	1 megaton
"Sark"	1959	300	1 megaton
"Serb"	1964	630	1 megaton
"Strela"	1961	400	kiloton range
"Scud"	1957	150	kiloton range
"Snaddock"	1962	250	kiloton range

through into international treaties almost entirely by the joint and exclusive efforts of the United States and the Soviet Union. The CCD is—and has been since it started as the Ten Nation Disarmament Conference—controlled by its Russian and American joint chairmen. They control its membership, its procedures, and, largely, the recommendations that issue from it to the U.N. General Assembly. This issue is beginning to be bitterly resented, both in the CCD itself and at the U.N. and the Draft Seabed Treaty, which purported to issue from the CCD, was returned to it by the General Assembly, after CCD member states had, in the General Assembly's Political Committee, tabled several amendments to it.

Such arms control measures as now exist are, in short, the fruit of political decisions taken by the Russian and American governments. There is nothing to suggest that technical questions have had any effect on the agreements they have come to, though they have had, of course, on the political decisions on each side. Political agreement between Russians and Americans is indeed a necessary condition for arms control, but that it is not sufficient is shown by the ineffectiveness and insufficiency of arms control up until now. The fact that SALT talks are now to start in earnest and that both participants are aware of the connection with the NPT, may allow those with an optimistic cast of mind to expect that, in future, arms control negotiations will take into consideration a wider range of interests than has so far been the case, and that progress towards actual disarmament is not impossible. Some further conclusions can be drawn from this study which may help us in considering arms control on the seabed.

CONDITIONS FOR SEA AND SEABED ARMS CONTROL

The term "Arms Control" is fundamentally unsatisfactory: the word control, in English, denotes restraint, the holding in or check, of this or that, but in French it denotes supervision, verification, inspection. The Russian translation approximates more closely the French meaning than the English, and from this ensues misunderstanding. However, in the case of the scarcely militarized and virtually unpoliced seabed, the word "disarmament" is clearly unsuitable, and "non-militarization" impossibly awkward as well as somewhat ambiguous; "arms control," in the English sense, of "holding in check" does seem suitable, and it is used in this sense. "Force Management" might be better, but "management" suggests a "manager" or "managers," and although international inspectorates and constabularies are already within the bounds of possibility, the international management of force probably is not.

To be viable, any arms control measures have to satisfy at least these conditions: 1) They must patently serve the interests of all those states whose governments' support and agreement is required. (This was not the case with the Non-Proliferation Treaty.) 2) They must be appropriate and applicable not only to the immediate situation and to the weapons systems actually being produced and deployed, but also to the future and to the weapons systems which are in the conceptual and research pipeline. (The Partial Test Ban was quite successful as a clean air measure, but because it has been possible to carry out nuclear tests very efficiently underground, it has had no effect on the central arms race until quite recently when it has somewhat constricted the development of ABM systems. Moreover, two non-signatories, France and China, have conducted atmospheric tests, without great international opprobrium.) 3) Means of verification must be tolerable, convincing and adjustable. (The Soviet Union at the present moment is not likely to find any on-site inspection tolerable. Neither superpower seems likely to submit itself substantially to any internationally controlled enforcement system. It is not yet clear that the NPT "safeguards," to be devised and operated by the IAEA over the civil nuclear programs of non-nuclear weapon states, will be judged tolerable by the latter unless similar safeguards are operated in all the nuclear weapon states. Equally, it is clear that verification arrangements in the Russian-American draft Seabed Treaty are not likely to be found convincing by other governments. How adjustable these various treaties may be will be seen when the time comes around for the review conferences.) Moreover, measures must be knave-proof as well as fool-proof. (Thus any arrangement about nuclear explosions for peaceful purposes in a Complete Test Ban will have to be devised so that the explosions cannot be used for the surreptitious testing of nuclear warheads.) 4) Preferably, for reasons of economy, convenience, and morale, the control should be part of ordinary on-going processes useful for other purposes as well. (The interweaving of measures to prevent the proliferation of nuclear weapons and to promote the peaceful uses of nuclear energy is the most promising element in the IAEA and Non-Proliferation Treaty set-up. Verification of a ban on underground tests may usefully be combined with an international seismic surveillance system, with resources exploration by satellite, and with an international system for the provision of nuclear explosions for peaceful purposes.) 5) Measures must be compatible with long-term requirements of arms control as part of a general international security system, since a measure which at one moment of time appears useful may in the long term fail to contribute to overall stability, or to the improved international system which it is the purpose of arms control to promote. (An instance is the Eisenhower Open-Skies proposal of 1955, which, had it been accepted, might have been briefly stabilizing but which shortly afterwards would have become highly destabilizing because of the development of long-range missiles and the valuable target information the "Open Skies" would have made available to a potential attacker. It is arguable that recent agitation in the United States against the Nixon Safeguard ABM project—as opposed to the Johnson "Sentinel" project—was misconceived because some element of missile-site ABM defenses may now be the only way of avoiding a large element of MIRV installment. This may be the only way of reconciling the superpowers to restraint while China remains outside any agreement; and also because it may, as Mr. Gromyko was first suggesting in 1963, be a useful, even a necessary, component of the transitional minimum deterrent framework of an eventual disarmament program.) 6) Lastly, the negotiation of an arms control measure must be both timely and sensitive. It is as highly political a process as any other in international relations: the product of competing and interacting pressures, insights and interests, within the international community, as well as within the individual members of that community, and it cannot be insulated from that normal context. (President Johnson's sudden decision in the summer of 1966 privately to negotiate a Non-Proliferation Treaty with the Soviet Union was an unusually abrupt and simple change of policy toward Europe generally and Western Germany in particular, and one effect of this was to make the powers inconveniently suspicious of a treaty in itself desirable. The way superpower attitudes to the SALT talks veered during 1968 and 1969, particularly those of the Soviet Union's during 1969, demonstrated very clearly the operation of internal pressures.)

If any useful seabed arms control is to be achieved, it will have to satisfy most or all of these conditions. It will also differ from other arms control arrangements in a number of ways arising out of the peculiar legal, economic and physical nature and status of the sea and the uses mankind makes of it. These, of course, are in turn constantly affected by new developments on land, in military and industrial technique and organization, in pollution control, alternative methods of protein procurement, and so on, and in the population explosion.

The differences between arms control relating to the sea and the seabed, and arms control in other elements derive from the following facts:

1. The absence of any basic arrangements for the maintenance of law and order, comparable to national police forces and specialized constabularies and inspectorates on land.

2. The principle of the freedom of the High Seas.

3. The fact that submarine-borne nuclear forces deployed at sea derive a large measure of invulnerability both from the nature and from the size of the element and are likely to continue to do so.

4. The absence or limited nature of national sovereign rights

not only beyond the Continental Shelf, but also on it, and in the High Seas with which, beyond the limit of territorial waters, it is covered.

5. Finally the fact that sea and seabed arms control, though no doubt generally held desirable, is not politically "hot" in comparison with previously negotiated measures. There is no public concern comparable to that over radioactive fallout which preceded the Partial Test Ban or over ABM and MIRV which preceded the Strategic Arms Limitation talks or even to that felt officially in Washington and Moscow over the horizontal proliferation of nuclear weapons which led to the Non-Proliferation Treaty. Any urgency the question of sea and seabed arms control may have arisen rather from the absence of any appropriately organized system of force management over a part of the globe in which mankind is progressively becoming more interested.

Let us examine these matters in further detail:

1. *The absence of basic arrangements for the maintenance of law and order.* To provide and protect that internal order which is called different things in different countries—in Britain it is called the "Queen's Peace"—is a fundamental duty of governments. The principal difference between the civilized condition and the wild is that in the former the exercise of force is brought to a minimum by the careful disposition of proper controls; in the latter it is free-lance and often ineffective and excessive. A modern government maintains a system of law and order of very great complexity (fiscal, industrial, social, planning both economic and physical, etc.), and implicitly coastal governments will be undertaking to provide all this on the seabed and in the sea above it whenever they undertake, or permit others to undertake, that exploitation of natural resources reserved to them in the 1958 Convention on the Continental Shelf, and also when they use these places for the disposal of waste. As far as the oil industry is concerned, where accidents can be of disastrous proportions, this may involve very substantial and expert inspection.[24] Moreover, because as interface between waters above and ground beneath, the seabed itself has no real existence; national or nationally licensed seabed concerns using the Continental Shelf beyond territorial waters, for any purposes whatever, will in practice

have to be policed within the inalienably nonnational waters of the High Seas. Conversely, coastal states will certainly seek to develop rights concerning activities on or in the High Seas which affect their interests on the seabed below as well as on their coasts.

It has sometimes been thought that the Antarctic Treaty and the Space Treaty, dealing as they do with other undeveloped and unpoliced areas, may provide useful parallels for sea and seabed arms control. This is unlikely to be the case: the economic exploitation of the South Pole and of Space are neither of them of great immediate promise, whereas both the sea and seabed undoubtedly are. The question of commercial enterprise is when, not whether. Although the military significance of space is likely to be important, there is a far higher technological threshold for the military use of space than for the military use of the sea. In space, for the time being, activity and verification of activity will be largely controlled by the United States and the Soviet Union.

Moreover ambiguities and ambivalences in the Space Treaty ("space" is nowhere defined in it; the Soviet Union's Orbital Bombardment System is not held to contravene it) makes it an unsuitable model for a treaty designed to shape and determine relations between a multitude of states and interests in that much and long-used element, the sea. Inspection on a basis of reciprocity is a concept full of pitfalls for the weak.

2. *The freedom of the High Seas.* This is in the process of diminishing: in the High Seas, as traffic increases in density, speed, and size of unit, and as the exploitation of natural resources also increases, there will be a parallel increase in regulations for the prevention of pollution, for the avoidance of accidents, for the rationalization of the fishing industry, for the conservation of resources, and so on. The development of *de facto* international constabularies and inspectorates will necessarily if slowly follow. The question of coordinating national and international bodies will arise and, in the case of the international bodies, the question of how they may acquire and operate the final sanction of force without which any policing arrangements remain ineffective. It is clear that for the time being, the provision of this final sanction is unlikely itself to be international.

There will also be the question of how underdeveloped countries with few spare resources of wealth or trained manpower will manage to service the national inspectorates and forces required to ensure the observance, not only of their own national regulations on the Continental Shelf by their licensees but also of internationally established rules, about, for instance, pollution control, for the observance of which they will be responsible.[25] The contractual shifting of responsibility from the national government of the coastal state to the exploiting companies now in practice frequently happens, but this and the use of arbitration as a kind of de facto solution to the juridical problem will not turn out to be a very satisfactory system in the long run. Because of this, small states may plausibly consider the setting up of cooperative regional policing arrangements which would have the advantage of requiring neither superpower participation nor Security Council agreement. Alternatively, organizations like IMCO might set up professional inspectorates and constabularies whose services small countries could call on. Eventually, it is perhaps not impossible that the Food and Agriculture Organization for instance should be provided with a fishery protection service of its own supported by naval units such as those the United Kingdom offered to the United Nations for "peacekeeping" purposes in 1965. Alternatively, a new Maritime Agency may be set up to group all existing sea and seabed related institutions and activities, and to perform whatever new functions it becomes necessary to assign to it. The failure of governments to provide policing arrangements while at the same time enacting regulations (e.g., insisting on the hydrocarbons industry and its carriers accepting liability for its accidents) could lead to private policing, run either by the industries concerned,[26] by the insurance world,[27] or by a kind of maritime securitor.

The freedom of the High Seas is clearly about to be eroded in a number of ways, particularly the untrammeled freedoms to carry and to discharge oil in them and to use them as a drain or rubbish dump. The dumping of radioactive matter is (or should be) already carried out in accordance with IAEA standards. Responsibility for scanning the North Sea for oil pollution and taking steps to control it when observed is now divided up among the North Sea coastal states.[28]

With the development of super-tankers of 400,000 tons and more, of dracones, of one hundred knot air cushion vehicles, some of the world's international narrows and other dangerous waters will see more or less strictly enforced traffic regulations. If these are not actually policed by way of governmental or intergovernmental cooperation,[29] the insurance world will come to do so indirectly by refusing to issue blanket policies,[30] or to accept liability when local "voluntary" rules are flouted.[31] Moreover, the conversion of fishing from a system of hunting to a rationally conducted major industry also cannot but diminish the freedom of the seas. Failing other methods, a vast extension of territorial waters, as declared by Chile, Peru, Ecuador and other Latin American states, and their defense, in some part of the world perhaps on regional lines, is not inconceivable.

3. *The strategic significance of the sea.* When we talk about reserving the sea and the seabed "beyond the limits of national jurisdiction exclusively for peaceful uses," we probably use the word peaceful as much in the sense of cooperative, agreed, not inducing controversy or dispute, as in the sense of totally demilitarized.[32] In May, 1969, Mr. Ignatieff (the Canadian delegate to the Geneva Disarmament Conference) quoted a view undoubtedly now shared by many seagirt states: "We do not interpret the phrase [exclusively for peaceful purposes] as prohibiting all military uses. While we strongly oppose military installations for offensive purposes, we have reservations about the desirability of precluding the use of the seabed adjacent to a coastal state for [its own] purely defensive purposes." [33] He went on to declare, "No other state may undertake these activities, or undertake any research on the Continental Shelf, without the consent of the coastal state. In discussion of this question elsewhere, some representatives have maintained that the jurisdiction of the coastal state does not extend to the freedom to install structures or devices which are unrelated to the 'exploration and exploitation of the natural resources' of the Continental Shelf—for example, purely defensive listening devices. It can readily be concluded, however, that states are not likely to ignore their security requirements simply because the Geneva Convention of 1958 on the Continental Shelf is silent or unclear on the subject. Moreover

. . . it is difficult to reconcile the coastal state's sovereign rights with freedom of military activity of any sort by foreign states on its Continental Shelf. Certainly Canada could not accept such activity on its Continental Shelf."

If this interpretation of a coastal state's rights on the Continental Shelf were generally accepted, and if at the same time the maximalist definition of the extent of the Continental Shelf is accepted by the principal oceanic powers, (i.e., that it extends to the outer edge of the Continental Slope, or even of the Rise) a substantial measure of seabed non-armament would exist, at least in declaratory form, together with enormous duties of verification falling on the governments of coastal states. However, the verification of what is going on on the seabed is probably an impossible task. "The maximum that we can project now in hunting for an object is a search rate for a given vehicle of about one square mile a day and there are 273 million square miles in the sea." [34] (It took over five months for the U.S. Navy to find the submarine *Scorpion*, after they had some idea of where to look for it.) Moreover, there is the high cost of such searching—a minimum of something between one-half million dollars and one million dollars per ship per year—that is for an area a mere thirty-six miles by ten miles.[35] This is a kind of expense that very few governments could contemplate, certainly not Canada, with three oceans to cover. Still, the possibility, even if slight, of having one's military equipment discovered and identified in an area purportedly clear of military equipment would have a marginally discouraging effect. Regional arrangements might in some places provide some assurance of nonmilitarization. Indeed constabularies which might be set up to oversee offshore installations licensed by coastal states for purposes of economic exploitation and pollution control could well have a useful function (in occordance with condition 5 above) [36] for verifying the nonarmament of the Continental Shelf. Those coastal states which could afford to, might prefer to perform this function themselves, but even so, substantial areas of the sea and the seabed would be subject to regular patrolling by a constabulary on the beat.

The Delegate of Malta at the U.N., Dr. Pardo, recently pointed out, speaking of the superpowers, that "these countries are very reluctant to commit themselves to an international regime [for the sea and the seabed] unless they are sure that it will maintain the present balance of power. This is a very important point, and no international regime with any administrative functions has any chance of acceptance unless this point is taken fully into account."[37] Because some kinds of demilitarization, for instance those proposed in the Soviet draft Seabed Treaty of March 1969, would offend against condition 1 (serving the national interests of necessary signatories), they are not seriously negotiable, nor, in consequence, for the time being worth advocating. It is not yet clear whether it will be possible in the near future to get beyond the rather indefinite undertaking the super-powers declare themselves willing to make in their Joint Draft Treaty "not to emplant or emplace . . . any objects with nuclear weapons or any other type of weapons of mass destruction. . . ."[38] This draft may perhaps rather be interpreted as something similar to their 1963 declaration, in which they undertook "to refrain from placing . . . objects carrying nuclear weapons or any other kinds of weapons of mass destruction" in orbit or anywhere in space, than as the substantial contribution to the new international regime that will eventually, because of their capability, be required of them. In fact the proposed ban on "emplacing or emplanting" nuclear and other weapons of mass destruction on the seabed might more plausibly have issued as a joint declaration from the SALT rather than be enshrined in an international treaty of as doubtful excellence as the seabed treaty currently in draft. A similar declaratory ban could concern nuclear and other weapons and weapons systems emplaced on artificial islands on the High Seas, whether over a Continental Shelf or not, and whatever the "island's" principal purpose.

For the time being at any rate the opacities and vastnesses of the deep sea contribute to the stability of the by no means overstable strategic balance and the laws of physics are such that this is likely to continue. Even if anti-submarine warfare techniques advance enormously in the next few years, the underwater cat-and-mouse game will remain subject to quite different calculations from over-water defense and this fact will itself have a bearing on the strategic balance.

Every increase in the accuracy of guidance systems for offensive weapons increases the vulnerability of the land-based systems on which they may be targeted. By the end of the seventies the kind of accuracy shown by the American moon shots may well be production-belt matter for the superpowers—hence the requirement for missile-site ABM defenses if retaliatory forces are not to become destabilizingly vulnerable. Consequently it is very unlikely that the existence and movements of ballistic-missile-launching submarines will come within the ambit of arms control discussion—although their numbers may—until the very end of a disarmament process. The apparently good news that issued from the preliminary SALT talks at Helsinki may make thinking about the "transitional minimum deterrent" relevant again, and if that is so, then it still looks as if a part of these minimum deterrent forces should be located under the sea. Their purpose would be to remain as inconspicuous and unobjectionable as possible, able to disappear at the first sign of other activity, constabular or economic or scientific. The proposed ban on "emplanting or emplacing" [39] nuclear-armed "objects" on or in the seabed is scarcely more than an admission that anything stationary is, in technical military terms, unattractive. If the ban, as Mr. Leonard, the U.S. representative, assured the CCD, covers vehicles that can crawl about the sea-bottom,[40] its importance would be rather more substantial. Crawlers and bottom resters might be cheaper for some purposes than the type of vehicle which would be designed not only to service them, but also to assist submarines in distress, inspect "objects," etc.: [41] once deep submersible technology has become "sweet," an observed ban on bottom-crawlers and even more on mobile bottom-resters, would certainly reduce the danger of a production breakthrough causing, as happened with Minuteman, a crisis in the strategic balance. Sea-based overkill could be as destabilizing as land-based. Whether the twelve mile coastal area exempted from the ban does not leave too much seabed unaffected needs careful examination. Dr. Craven's view is that anything more than one mile "would make a complete prohibition in the rest of the world meaningless." [42] In any eventual seabed treaty, what precisely is meant should be spelled out and ambiguous words like "emplace" and "emplant" be avoided.

There is one further small sense in which total demilitarization of the seabed is unlikely: life insurance for civilian crews of deep-diving submarines is so expensive that it makes the use of service personnel for exploration comparatively attractive. This use of service personnel is specifically allowed for in the Space Treaty, the Antarctic Treaty and the draft Seabed Treaty.

4. *The absence or limited nature of national sovereign rights.* This fact, combined with the physical shape of the underwater, makes it certain that in this element, warfare (other, that is, than missile launching) will more closely resemble guerilla warfare than any other. When Chairman Mao speaks of the guerilla fighter being "like a fish in the sea" and of "the ocean of a people's war," he is saying something which, inverted, is still instructive. The sea is a no-man's-land, where uncertainty and solitude will prevail for a long time. As far as the underwater is concerned, below a certain depth there can be no confidence in the adequacy of verification procedures for any measures of arms control. On the other hand, there can be no problems with the fear and accusation of espionage: who can inspect, may. But most governments would find this a slender foundation on which to base the security of their countries and much of the verification of seabed arms control will have to take place elsewhere than at sea, that is, in shipyards, harbors, factories, inventories, etc., where, of course, the usual difficulties obtain. As the seabed is used and exploited and patrolled it will to that extent lose its jungle character: where it is civilized there may well be effective arms control by means of various interlocking national and international constabularies and inspectorates. In the wild redoubts, there will be neither kind of control because there cannot be, and the rest of the international regime will have to be so devised that the redoubts do not signify too much.

NEVILLE BROWN

20 Military Uses of the Ocean Floor

As the history of the inter-war period shows, arms limitation on the sea surface tends to be considerably easier than it is on land. To build a large number of warships invariably takes a long time; and, when they have been built, they are easy to identify and hard to conceal. Unfortunately, however, the sea-floor does lend itself relatively well to clandestine exploitation for military as well as for other purposes. This is why it is so desirable that an arms control regime be imposed on this surface before the militarization of it has proceeded too far.

In any examination of the prospects for such a regime, the most important topographical distinction to preserve is that between the Continental Shelf and the Deep Seabed. Just how crucial this is, from the military point of view, can readily be seen from the history of the Second World War. For the Shelf and its margins often constrained vertical motion beneath the sea surface and also acted as a stable base for mine-anchors and other installations. For example, the defensive mine barrage laid on the submarine ridge between Scotland and Iceland and that laid off the East Coast of England helped to inhibit the movement of U-Boats. So

---

NEVILLE BROWN is a Lecturer in International Politics at the University of Birmingham. He also writes regularly on defense for the *New Scientist*. He has served as a meteorological officer in the Royal Navy and has been on the staffs of the Institute for Strategic Studies and the Royal Military Academy, Sandhurst. He is the author of several books and reports on strategic studies and contemporary history. Among them are *Strategic Mobility* and *British Arms and Strategy: 1970–80.*

did the shallowness of such sea areas as the English and Bristol Channels. Testimony to the threat posed by the defensive mine to intruding surface vessels can be seen in the employment of scores or even hundreds of minesweepers in support of every large British and American amphibious operation in the European and Pacific theatres. But mines figured no less prominently in offensive action. Both the British and the Germans laid well over one hundred thousand in European waters: each sank over one thousand enemy vessels by so doing. Sometimes moored mines were laid above seabeds as deep as one thousand fathoms. But usually they were laid above the Continental Shelf.[1]

Boom defense nets were widely used to provide vital anchorages with local protection against submarine attack, Britain alone having seventy-five boom defense vessels in service by 1945.[2] Then again, the seaward approaches to, for instance, London were made more secure by small concrete forts mounted on piles (and outside territorial limits). The oil pipeline laid from England to the Normandy beach-head in 1944 affords another example of the exploitation of a shallow sea-floor. So does the use, admittedly limited, made of underwater beacons to mark fairways through mined areas. A further consideration is that enemy telegraphic cables could be severed on the Continental Shelf but not, as a rule, beyond it. And to all the direct and specific implications of the presence of this Shelf must be added the way in which passage into the shallower waters always caused both tides and wind-driven waves to increase in amplitude.

All these aspects remain technically germane and to them can be added the interest now being shown in the Shelf as a platform for surveillance. Therefore, although much scope exists for speculation as to how widespread international conflict will be in the years ahead and what forms it will take, it may be instructive to consider which strategically significant stretches of water lie above the Continental Shelf. The 1958 Convention put the general limit of the Shelf at a depth of two hundred meters, whereas many atlases put it at one hundred fathoms. In fact, however, these two submarine contours are almost coincident, two meters being about 1.09 fathoms.

One major section of the Shelf underlies the Baltic, all of the

North Sea apart from the Skaggerak, and the Western approaches to the British Isles. Also, the submarine ridges between Scotland and Iceland can be regarded, at least from some operational standpoints, as part of the same zone. Similar stretches off Southern Asia include the Persian Gulf, practically all the South China Sea and much of the water embraced by the Indonesian Archipelago. Around the Eastern and Western seaboards of North America the width of the Shelf is generally between twenty and 120 miles, but it tends to be much broader on the polar flank. Thus, the fact that Hudson's Bay is both shallow and hemmed in by an extensive island chain makes of little account what might otherwise be a profound weakness in the strategic geography of Canada. Another feature that could have major naval implications under certain circumstances is the broad stretch of shallow water between the Arctic and the Pacific. About seven percent of the sea surface of the world covers a floor less than one hundred fathoms deep and ten percent a floor between one hundred and one thousand fathoms down. The average depth of the Open Sea is 1,900 fathoms.

Both active and passive sonar, i.e. sounding, devices have been in use in many navies for some time. The former pulse their own acoustic energy and so can calculate distance from the time an echo takes to return. Passive sonar, being dependent upon the reception of noise from the target itself, affords no direct measure of range. Another weakness inherent in passive sets is that they usually have to rely on a broad frequency band, a characteristic that renders them more susceptible to being either misled or deafened by spurious emissions and also to being deliberately jammed. Perhaps the chief drawback of active sonar is the one that derives from the basic geometry of wave motion. With sound, as with all forms of radiant energy, the recorded strength of a reflected pulse varies inversely with the fourth power of the range being achieved, e.g., when this is doubled the echo received from a given target diminishes by fifteen-sixteenths.

Nevertheless, standard sonar sets of the active variety mounted in either surface vessels or submerged submarines can very occasionally recapture pulses that have bounced off targets well over fifty miles away. Furthermore, passive sensoring equipment located in such vessels has been known to pick up emissions across one hundred miles or more. Even so, such distances are puny in relation to the total extent of the oceans and do not, in any case, represent norms. Mechanical and biological sea noise will often cause serious interference. So will the self-noise a vessel traveling at speed is liable to generate. Above all sound waves will be extensively reflected and refracted by the variations in temperature and salinity that can occur in sea-water, especially in the vertical plane; and this reflection and refraction hampers acoustic detection in the majority of cases. The net result is that the average range of first sonar contact from ship or submarine is still only of the order of five or ten miles.[3]

Hence the attention being paid, at least in the United States, to the possibility of building exceptionally large sonar sets and installing them on the Continental Shelf and, perhaps more particularly, on its "high ground" or near its seaward edge. For this kind of location offers distinct technical advantages. The efficiency of the sets will be little impaired by the very marked changes, either gradual or abrupt, that can occur within several hundred feet of the surface. The water at lower levels tends also to be a good deal less turbulent. Submarines often present more conspicuous silhouettes to sensors localed below them. Then again, more bulky sets may be installed on the seabed than any ship could accommodate. Obviously this means higher outputs of power can be achieved. What also is implied is that greater ranges can be obtained through the use of lower frequencies. For, below fifty kilocycles per second or thereabouts, the absorption of sound in sea-water is proportional to the square of the frequency. But one of the complications is that to halve the median frequency of a pulse of given strength requires a sonar set of about eight times the weight and volume; and, beyond a certain point, this means emplacement on the sea floor.

In fact, throughout the last decade[4] reports have been appearing of sonar sets with maximum ranges of a few hundred miles being in position on the sea floor in certain strategic areas. Project Artemis, a program the U.S. Navy has for some years been engaged in, is especially concerned with VLF sonar transducers. Some of these are believed to be designed for the floor while others are borne on surface ships. Among the latter is the experimental model that extends through several decks of the seventeen thousand ton converted tanker, Mission Capistrano. Presumably some of the floor-based models are of comparable size.

Late in 1969 it was reported that the United States Navy had begun further to strengthen its arrangements for the offshore passive detection of submarines at considerable depths and extended ranges by the installation, on an experimental basis, of a network of hydrophone arrays known as Sea Spider. Unlike the Project Caesar already in position along parts of the Atlantic Coast, "Sea Spider" will be powered by self-contained nuclear batteries rather than shore-based generators. All the sea-based systems so far developed are unmanned. Nevertheless, a long-term possibility does exist that permanent manning will prove desirable for monitoring and data-processing.

What then of the other naval applications of the Shelf? Underwater sound beacons used in conjunction with sonar sets installed in submarines may help the latter get fixes sufficiently accurate for, let us say, the firing of Polaris or Poseidon missiles against relatively hard targets. Any such beacon would probably be activated by pulses emitted by a submarine's transducer on a predetermined frequency. But devices of this kind could serve as aids to merchant vessels as well. Indeed, their potential usefulness may lie mainly in facilitating the passage of merchantmen through narrow channels.

Mines that depend for their activation on physical contact must, of course, either be moored or else left to float freely.[5] Conversely those which depend on the fluctuations in hydrostatic pressure caused by passing ships are always "ground" mines, i.e., they rest on the ocean bed. Those activated by magnetic or acoustic fields (or by a combination of the two) may be either ground or moored. Emplacement of the former sort is subject to quite severe constraints, thirty fathoms being about the maximum effective depth for "depression," i.e., hydrostatic, mines and not much more for those which respond to magnetic or acoustic "influence." Within their limits, however, depression mines, in particular, are a considerable menace. For they are extremely hard to find and neutralize, especially when lying on a coarse and rugged surface. Part of the trouble is that no such device will be more than several feet across, which means that even high definition sonar would be unlikely to detect them from more than one or two hundred feet away.

All the same, the implications for verified arms control are not so grim as they might appear. For one thing every naval power can feel confident that, should a military crisis arise, defensive minefields could be laid very quickly. Thus, a submarine or destroyer adapted to minelaying might well be able to carry sixty mines in a single lift; and figures given in Jane's *Fighting Ships* [6] indicate that the Soviet Navy, for instance, could lift simultaneously well over fifteen thousand. Nor should figures like this be dismissed as entirely notional. At the beginning of World War Two barrages consisting of thousands of mines were sometimes laid in a few days. This means that little would have been gained from laying them prior to the outbreak of hostilities.

The fact that defensive mines need not be laid very long before they are required has already had a beneficial effect on the evolution of the relevant parts of the law of the sea. For this now categorically stipulates that neither mines nor anything else may be laid in international waters in time of peace in such a way as to endanger shipping. Under all circumstances, moreover, any nation laying mines is expected to issue at least general warnings to this effect.

In the early 1960's some consideration was given in both the United States Air Force and the United States Navy to the possibility of putting intercontinental rockets in containers that could be moored, fully submerged, in shallow waters. Recognition has become widespread, however, that such systems would be unattractive in that they would lack not only the protection of "hardened" sites that weapons like Minuteman enjoy but also that which continual mobility confers on the Polaris boats and other vessels of a similar kind. In addition, the apparent need to resort to compressed air to thrust a rocket above the sea-surface would probably mean it would have to be situated less than fifty feet beneath that surface; the difficulties of remote command and control by means of wireless waves might impose a similar constraint. Admittedly, weapons like Minuteman may soon become a great deal more vulnerable to surprise attacks on their launching sites as a result of the introduction of Multiple Independently-Targetable Re-entry Vehicles (MIRV's), multiple warheads that spread out from the capsule of an offensive rocket as it is coming back to earth and then proceed to their targets on individual and very accurate trajectories. But an increase in ballistic missile submarine forces and, in due course, the development of mobile land-based ICBM's are likely to prove far more attractive solutions to the problems thereby posed than systems tethered to, or buried in, the sea floor.

What about the deployment by the two Superpowers of Anti-Ballistic Missiles (ABM's) either to protect the sites upon which strategic retaliatory forces are based or else to defend the whole of their respective national territories? Apprehension about the MIRV threat has already induced the American government to start to provide Intercontinental Ballistic Missile bases with point-defense by means of medium- and short-range ABM's. Furthermore, it has led some analysts to conclude that what the U.S.A. really needs are ABM's able to make interceptions not when the flurries of multiple warheads are descending towards their targets but well before warhead separation has occurred. Among the inferences that can be drawn from this proposition is that a substantial part of the ABM force should be situated at sea.

What configuration would a sea-based ABM force have? Would technical considerations favor its being mounted on floating platforms anchored to features such as the Mid-Atlantic Ridge or would it be better shipborne? Being able to maneuver close to an adversary's coastline, ships could conceivably watch at least a proportion of his ICBM's during their initial ascent, a phase in which they are highly susceptible to tracking. Among the other advantages mobility could confer are more immunity from harassment and more scope for demonstrative posturing in time of crisis. A point to bear in mind, on the other hand, is that the radar performance an ABM system would call for would exceed in scale and subtlety that so far achieved by any ship. Even so, it is the opinion of at least one informed student of this subject that recent developments in the radar field have made this problem manageable.[7]

Another possibility to reckon with is that eventually an anxiety to provide all-round ABM defense will generate an interest in oceanic locations not for complete ABM networks but for some of the perimeter radars that may give the earliest warning of a missile attack and the first indications of its scale and character. Yet in this case, too, it may be doubted whether any systems would have to be mounted on tethered platforms. Some combination of ships and islands would probably constitute at least as suitable an environment.

Therefore, ABM elements based at sea—however controversial they may prove for other reasons—need not be expected directly to undermine the international and nonmilitary status of the mid-Atlantic ridge or any other part of the ocean floor. But there is one way in which extensive ABM deployment could compromise attempts to avert the militarization of this floor. Let us suppose, to take a perfectly plausible example, the U.S.A. eventually creates a nationwide ABM coverage against the relatively light and primitive missile attack that could conceivably be launched by China in, say, the early 1980's; let us assume also that, whilst she does so, China reinforces her land-based rockets, with a number of ballistic missile submarines. Would not the logic of the situation then be that the U.S.A. would have to establish virtually impenetrable Anti-Submarine Warfare (ASW) barriers along its seaboards? Otherwise, surely, these Chinese submarines could simply move close inshore and fire their rockets at coastal cities on trajectories so short and flat that no ABM batteries would be able to engage them.

Presumably these ASW screens would largely consist of ships, submarines, and aircraft. No doubt, too, there would be lines of seabed sonars. What can be discounted, however, is that anybody would wish to add to this list sea-mines with nuclear warheads. Two justifications for this assertion can be advanced. One is that, even if every single mine was controlled by its own electric cable, there would be an unacceptable risk of accidental detonation. The other is that, from the tactical standpoint alone, the mine is not at all satisfactory as a nuclear delivery vehicle. Thus, a nuclear warhead of, say, a high kiloton yield would typically be able to wreck a submarine several miles away: in other words it would have a kill radius well in excess of the maximum distance at which it could itself be activated by the passage of a vessel. Therefore, any mines incorporated in an anti-submarine barrier—even one intended to thwart the Chinese equivalent of a Polaris boat—would almost certainly contain charges composed merely of chemical high explosive.

But the fact that weapons of mass-destruction would probably be excluded does not mean that a barrier of this sort would have no adverse implications for arms control. To be at all effective such a screen would have to extend its surveillance at least one hundred miles offshore and be ready to come to full operational status the moment a risk of nuclear war developed; and by full operation is meant a situation in which every vessel suspected of being a hostile ballistic submarine was being sunk on detection. Yet even a threat to use an ASW barrier in this manner would be manifestly incompatible with the traditional freedom of navigation of the High Seas in time of peace. Therefore, it is hard to imagine sensors placed on the sea-floor as part and parcel of an endeavor to back up an ABM area defense not being widely regarded as contrary to the spirit of maritime arms limitation.

Nor does the solution lie in precluding barriers of this kind by going beyond the terms of the present draft treaty and banning from the sea-floor warlike structures of every size and description. At least three objections can be leveled at this. The most ineluctable is that the control of mines and the smaller types of listening devices would be next to impossible to verify. The second is that the outcome would be asymmetrical in that it would place at a disadvantage those nations with both the opportunity and the necessity to use the Continental Shelf as a giant observation platform in order to protect their legitimate maritime commerce. The last point, and quite an important one, is that a few sensors in the fairways leading to and from their own bases could help preserve the members of a ballistic missile submarine fleet from destruction. This is because, by keeping watch on adversary movements, sensors can minimize the risk of friendly boats being tracked as they commence or terminate a patrol. Nobody imagines that this sort of tracking is very easy. Nevertheless, it is just about conceivable that it could serve to bring about the wholesale elimination of a small ballistic missile submarine force or the gradual attrition of a large one.

The truth of the matter may be that, as appears to be the case with certain other aspects of the arms control debate, the key to any real breakthrough must lie not in the particular subject-matter under discussion but in a stabilization of what is sometimes spoken of as the central arms balance—the deterrent relationships between the major nuclear powers. For if, in particular, the two superpowers can be persuaded to abstain from the deployment of ABMs on an area basis the outlook is good for the militarization of the seabed being satisfactorily limited during the next decade or two. Otherwise it is not.

Perhaps it is desirable to consider in conclusion whether the whole issue will eventually become hopelessly confounded in consequence of armed protection being accorded programs for the peaceful exploration and exploitation of the sea-floor. Happily the risk of this seems minimal for as far ahead as one can reasonably hope to foresee. Very little interest has been shown as yet in this particular security requirement. Besides, for many years to come the pursuits referred to are likely to be concentrated predominantly on the Shelf, i.e., sufficiently close to the surface to be safeguarded by traditional naval methods. Weapons systems that can operate at great depth may, of course, be called for later on. However, these are much more likely to be advanced forms of submarine than structures that move on, or are fixed to, the ocean floor itself.

ROBIN MURRAY and FRANCES MURRAY

21 An Examination of the Existing Constabularies and Inspectorates Concerning Themselves with the Sea and the Seabed

This chapter is concerned with the ways in which rules, regulations and laws governing existing uses of the sea and the seabed are currently inspected and policed. By discussing present practice in nonmilitary marine superintendence, we hope to throw light on certain of the political, economic, and technological problems which are likely to arise in the implementation of verification measures for disarmament agreements on the sea and seabed. Further, in designing such disarmament verification measures, it may be possible to make use of existing forces exercising a constabulary function in nonmilitary marine matters, since from an economic point of view, existing police forces and inspectorates are able to extend their functions in a given area at a low marginal cost. The current role of national navies in nonmilitary policing at sea is an illustration of this.

What follows has been written with these points in mind, but we have purposely left it to those directly concerned with marine disarmament provisions to make explicit the implications from the nonmilitary discussion. We have confined ourselves to outlining what may be called the hardware of nonmilitary marine constab-

FRANCES MURRAY read Modern History at Somerville College, Oxford. Since then she has been principally employed as a secondary school teacher in London, specializing in remedial reading and the teaching of social science.

ROBIN MURRAY read Modern History at Balliol College, Oxford and then took a Masters Degree in Economics at the London School of Economics. He taught for two years as an adult education lecturer, and since 1966 has been a lecturer in economics at the London Business School.

ulary forces: the vessels, instruments and people used to enforce marine laws and regulations. The main nonmilitary uses of the sea with which we will be concerned are: fishing and mariculture, shipping, pollution, raw material exploration and development, submarine cables and pipelines, ocean data collection, broadcasting, and the infringements of national boundary legislation concerning customs, migration, and exchange control. The discussion will be concerned with territorial waters and other areas of the sea under national jurisdiction as well as with the high seas, since it is in the former that policing has been most fully developed.

FISHING

With a few exceptions, the policing functions of laws and regulations on fishing are performed by national forces. These forces have a double task: first of preserving the rights of their countries' fishermen when these are challenged by foreign fishermen or other sea users and second of ensuring observance of laws and regulations by their own countries' fishermen.

1. *Constabulary forces for surveillance within national fishing limits.* Intra-limit fishing constabularies will clearly tend to vary according to the importance which a country attaches to maintaining intact her exclusive fishery limits. The failure to adopt standardized fishery limits was one of the features of the 1958 Geneva Conference, and the fact that the problem is still unsolved means that there remain widely divergent claims. A recent survey of 119 countries, indicates that of the sixty-six with known fishing limits, eight claimed less than twelve miles, thirty-eight claimed twelve miles, and fourteen claimed more than twelve miles, twelve of these being claims exceeding one hundred miles (see Appendix).[1] Given these divergencies, it remains true that most national constabularies are composed of patrol vessels plus onshore inspectorates, backed up in certain cases with aircraft reconnaissance. These vessels may specialize in fishery patrol, or be seconded from the national navy, or again be part of the national navy fulfilling the fishery patrol function at the same time as undertaking more general duties.

The U.K. fishery patrol squadron is concerned above all with the prevention of unauthorized intrusion by foreign fishing vessels in the twelve mile limit. Within the three mile limit, local fishery committees have the responsibility for implementing national fishery legislation and local bylaws, and some of these run patrol vessels. In Scotland the Department of Agriculture and Fisheries for Scotland run a fleet of eight vessels for the superintendence of inshore fisheries along the Scottish coastline, whose activity has been principally concerned with illegal trawling and seining.

There is, too, an onshore fishery inspectorate consisting, in England and Wales, of twenty inspectors and officers whose job is to inspect equipment, make spot checks of fish sizes in the markets of major ports and also act as a fishery intelligence service.

The U.K. fishery constabulary thus appears far from homogeneous. The fishery patrol squadron is under the authority of the Ministry of Defense, as are the helicopters which are occasionally called out to survey suspected foreign poachers—though these helicopters may come from either the Army or Navy. The inspectorate is under the Ministry of Agriculture, which is also the prosecuting body for infringements. Finally the Department of Agriculture and Fisheries in Scotland has its own superintendence fleet and legal powers.

We may note, however, that this division of powers does to some extent reflect the distinction in the function of fishing constabularies between the protection of the fishing rights of one's own nationals against foreign fishermen and the ensuring of observance of fishing laws and regulations by one's own nationals. In the latter case, national land constabularies may be relied on for the seizure and arrest of infringers, whereas in the former case this is not so unless the foreign vessel puts in to the aggrieved country's port. Thus fishery patrols against foreign poaching have required the power, including fire-power, to arrest foreign vessels and this has often meant that naval ships are used, under the control of the military, for this purpose.[2]

2. *Constabulary forces for surveillance outside agreed fishing limits.* The distinction of the last paragraph is operative also on the High Seas. Here the protection of the rights of one's own na-

tional fishermen is in terms of ensuring their freedom to fish in certain waters, rather than the exclusion of foreign vessels. Britain again supplies an interesting example during her dispute with Iceland between 1958 and 1961. In September, 1958, Iceland extended her exclusive fishing limits to twelve miles, an act which Britain claimed unfairly kept out British fishermen from traditional grounds. During the dispute Britain sent warships to support her fishing fleets against the armed Icelandic patrol vessels, and though the warships never in fact fired, there were fourteen cases where they prevented the Icelandic patrol boats from arresting British trawlers fishing within the twelve mile limit, either by ramming them or threatening to sink them.[3] More generally, British frigates will accompany a British fishing fleet *en passant* for what officials refer to as "moral support."

The United States have defended, or attempted to defend, their fishermen's rights on the high seas in a less direct way. They have made a practice of lending naval vessels to foreign countries, but these loans are immediately terminated if the country to whom the loan was made is found to have seized a U.S. fishing vessel because it was fishing in what the U.S. recognizes as international waters.[4]

In assessing the possible forms of high seas fishery policing two variables stand out as particularly significant: first, the seriousness of the conservation problem as it affects all parties; and second, the cost and administrative character of the regulations envisaged. The first will in part determine the willingness of fishery states to forego some part of their national sovereignty for the sake of their economies. The second will be relevant when, for example, a convention's regulations cannot be adequately enforced from onshore. Where enforcement requires a seaborne inspectorate, costs rapidly become prohibitive not only for areas far removed from member states (the Convention area of I.C.N.A.F., for instance, is a considerable distance from the majority of its members in Western Europe) but even for wealthy and proximate states. The International Pacific Halibut Commission, whose members are Canada and the U.S.A., has several times referred to the disproportionate expense that effective control involved.[5] Given economies of scale in the policing function, certainly as far as the verification of offenses is concerned, and given the difficulties of enforcement from shore, we may expect economic pressures at least to support the development of "co-operative" if not "integrated" enforcement systems.

Our general conclusion is that the type of constabulary force required for fisheries differs according to whether the policing is of laws and regulations in defense of the rights of the nation's own fishermen or whether it is concerned with ensuring their good conduct. It also varies in relation to the extent that the laws and regulations affecting the high seas are seriously supported by the countries directly concerned with the fishing of those high seas.

MARICULTURE

The transferring of highly valued fish to good high seas feeding areas, as well as other forms of fish farming, awaits not only a regime of law, but perhaps more importantly a constabulary force capable of implementing law. Mariculture has been restricted largely to coastal waters, where law holds and policing is less costly. In Britain, for example, where mariculture is concentrated on oysters, mussels and clams, though on a scale far smaller than that, say, of Australia, there is no special policing of the areas of cultivation. The constabulary function is performed by land-based police forces and anyone found taking shellfish from these areas is prosecuted under the normal laws of larceny.[6]

SHIPPING

Laws and regulations applying to shipping are principally concerned with securing the rights of innocent passage and freedom of the high seas, with preventing and mitigating accidents and loss at sea, and with preserving common law and order on board ships.

In the case of innocent passage and freedom of the high seas, those vessels meeting with what they consider to be a restriction of their rights may be relied upon to inform their flag states of possible infringements of international law and convention rulings. The constabulary function will then be less concerned with the problem of establishing that an offense has been committed than with reestablishing the right once this has been challenged. This it may do either with the force of arms or through diplomatic channels. In the recent case where an American charter ship searching for sunken Spanish gold was seized by a Cuban patrol boat fifteen miles off Cuba, the U.S. government secured the release of the ship through diplomatic means.[7] Most of the restrictions of passage on the high seas derive from military considerations, such as the Cuban blockade, the Beira blockade or the capture of the "Pueblo," and, therefore, fall outside the scope of this paper.

The prevention and mitigation of accidents and losses at sea covers three sets of laws and regulations: those concerned with the seaworthiness of ships, their proper equipping, and their competent handling, those concerned with the provision of information to sailors about natural or man-made hazards, and those concerned with "rules of the road" to be followed by vessels for their own and others' safety. Where these are mandatory, they are almost all enforced from on land.

In the case of the seaworthiness of ships, for example, the great majority of merchant ships will be subject to frequent surveys by Classification Societies, whose assessment will form the basis for the establishment of premiums by insurance companies. Nonconformity to internationally accepted standards of seaworthiness will be penalized, informally, by the market. Further, all significant flag states will, with greater and lesser degrees of substance, make the registration of a ship and the granting of necessary certificates and licenses dependent on the achievement of certain minimum standards with regard to the hull and basic construction (including subdivisions and stabilization), machinery and electrical equipment, fire protection and precautions, life saving appliances, lights, radio equipment, and so on.[8] In Britain surveys of safety standards are carried out by Board of Trade surveyors, and in certain instances by surveyors of Classification Societies to whom power to carry out these statutory functions has been delegated by the Board.

Whereas there is an elaborate code of laws and repeated inspections as far as the basic hull of a ship and its equipment are concerned, the same is not true of the manning of ships. Those in charge of ships are required to have Master Mariners Certificates, and there are of course in most countries extensive provisions about pilotage and the qualifications required by pilots. But requirements for the achievement of qualifications vary widely. In some countries, Masters Certificates of Competency can be achieved without examination. In Britain where an officer will usually obtain his Masters Certificate around the age of twenty-five, there is no re-checking or re-licensing as there is in the case of airline pilots. Nor are refresher courses required, even for those who have spent some time away from the sea. When an officer wrecks a ship only the flag state has the right to investigate, and the flags of convenience (with now over half the world's shipping under their flags) are reportedly somewhat lax in their investigations. Further, while the Panamanian and Liberian governments may remove their own certificate which they have granted to an officer, the fact that these are granted without examination to anyone holding a certificate of another nation, means that such an officer would continue to hold a Mariners Certificate even after being found responsible for bad navigation. Even the insurance companies require no extensive details of the master and watch-keeping officers, going rather on the owner's record. Thus, not only are the provisions regarding the officer manning of ships somewhat rudimentary; so, too, is the exercise of what inspectorate or sanctions-imposing functions there are (withdrawal of license or raising of premiums).

We turn now to those laws and regulations concerned to provide information to sailors about natural or man-made hazards. The provision of such information is often the responsibility of the public authority who might be expected to police it. The constabulary function will in this case be in the nature of intra-organizational control. In the U.K. the extensive system of lighthouses, lightships, markings of shipping lanes and of wrecks, as well as lighted and unlighted buoys is supervised and inspected by three lighthouse authorities. Trinity House, the lighthouse and pilotage authority for England and Wales, has one to two inspectors who inspect navigational markings every three years.[9] In the provision of notification services such as updated information on markings, changes of sea level, movement of buoys and so on, or the sustaining of broadcasts of meteorological information, including weather forecasts, control again tends to be intra-organizational, concerned with the standard rather than the actual existence of the service.[10] Lastly, where the information is provided by ships themselves, in the form of lights or broadcast messages, enforcement is carried out partly by inspection of the equipment on shore and partly through civil actions and the evidence of plaintiffs who claim to have suffered from the faulty provisions of such forms of information.

Civil actions and plaintiffs' evidence are likewise the effective constabulary as regards most rules of the road. In the vicinity of ports, lanes are likely to be mandatory, but most other shipping lanes are not so. The North Atlantic Lane Routes Agreement was an agreement between the large private liners using the route, with the informal backing of the contracting governments of the International Convention for the Safety of Life at Sea. Companies were required to give public notice of the regular routes they proposed their ships would follow, and the ice patrol service, managed by the U.S.A., was required to report to the administration concerned, any passenger ship observed not to be on any regular, recognized or advertised route, to be crossing the Newfoundland fishing grounds during the fishing season, or to be passing through regions believed to be endangered by ice. But the agreement remained a private one and has recently been abrogated after being undermined by the fact that many other liners and all cargo vessels retained and implemented their freedom to use their own selected tracks.[11]

Let us turn finally to the third set of laws and regulations applying to shipping, namely those concerned with the preservation of social as against navigational law and order at sea. In this category we would specify among other things, those codes dealing with the safety, welfare and working conditions of the crew, and with piracy.[12] In the case of the working and living conditions of the crew, certain means of enforcement are specified in the international conventions on the subject. Inspectorates exist with parallel functions to the factory and health inspectorates of land-based activities. The I.L.O. Recommendation Concerning the

General Principles for the Inspection of the Conditions of Work of Seamen (1926) specifies that inspectors should be empowered to prohibit a boat leaving port until it conforms to specified legal standards, and that both the master of a vessel or members of the crew be entitled to call for an inspection.[13] The crew has similar rights to call for an inspector with respect to food and catering standards under the 1946 Food and Catering for Crews on Board Ship Convention.[14] Thus information about contraventions of these codes can be expected to come from (a) members of the ship, suffering from the contravention, and (b) land-based inspectors. Sanctions in the form of the delay of the ship and of the meeting of minimum requirements are in the hands not only of local courts but in some instances of inspectors themselves.

We have seen that in the three sets of laws and regulations regarding shipping, there are few instances of seaborne inspectorates. Information about contravention is either derived from those suffering from the contravention, or by land-based inspection of conditions and equipment. Contraveners face sanction either by virtue of the contravention itself (using the wrong channel in a strait, for example, or possessing an unseaworthy ship) by civil action against them, or by public prosecution. Such a system of enforcement derives not only from the nature of many of the shipping regulations (it is in the common interest of mariners to observe them) but also from the common agreement among maritime states about the laws of shipping and enforcement systems which prevent wrong-doers escaping into jurisdictional vacuums.

POLLUTION

One of the main criticisms of the 1954 International Convention for the Prevention of Pollution of the Sea by Oil, as amended in 1962, which is the basis for laws concerning oil pollution on the High Seas, is the difficulty of enforcing its provisions. Most classes of ships are required to keep a log book of oil discharges and losses, which is a form of auto-policing, though this requirement itself has to be policed. The provisions for inspecting the log books vary considerably: France and Belgium for example, have a reasonably thorough inspection mechanism. Belgian ships have to send extracts from their log books (which would include their oil log) to the Tribunal de Commerce when they return to a home port, or to the Belgian Consul when in a foreign port. Other countries, particularly some smaller ones, are reportedly somewhat reluctant to have a stringent inspection scheme for fear of frightening away ships who might otherwise use their ports. The inspection of oil log-books is land-based.

The de facto constabularies for high seas pollution are clearly inadequate to enforce the laws as they now stand. The information they provide has not generally been a sufficient basis on which to mount successful prosecutions. The mere reporting of ships and the warnings issued by governments to owners where no prosecution occurred, are themselves a form of sanction. Within territorial waters the problem of enforcement is somewhat easier since conditions of proof tend to be less stringent and countries have a greater authority over foreign ships.

Many countries prohibit the discharge of oily mixture completely within their territorial waters, and define oily mixture more catholicly than is customary for the prohibited zones on the High Seas. Some, too, have constabulary forces over and above those mentioned in the previous paragraph. The U.S. Coast Guard have mobile cutters to track infringements. Japan has patrol boats on the alert in areas where oil discharges are likely to occur. Canada uses patrol boats and helicopters on the St. Lawrence River. Poland has a net of twenty-four permanent control stations along the whole of her coastline. We should also note that many countries have inspectors to check regulations designed to prevent oil pollution of the sea; oily-water separators, facilities for disposal of waste at oil terminals, and so on.[15]

Greater restrictions, more extensive detection measures and the right of jurisdiction over foreign ships in territorial waters have all made for higher rates of prosecution and convictions for offenses within territorial waters as compared with those committed outside. Canada reports a high proportion of successful prosecutions against offenders. Romania has on a number of occasions prosecuted and fined foreign ships which have discharged oily mixtures in the Romanian coastal zone of the Black Sea. In the case of the U.K. the figures for prosecutions compare interestingly with those given for extra-territorial sea offenses.[16] In 1967, out of sixty-two prosecutions there were fifty-nine convictions, including twenty-one U.K. ships, thirty-three under foreign flags, and five land installations; in 1968, out of sixty-four prosecutions there were sixty-two convictions, twenty-seven being U.K. ships, thirty sailing under foreign flags, and five land installations.[17]

The information about effected or potential oil pollution at sea is derived from public and private vessels and aircraft away from land, backed up by on-land inspectorates. Most of these bodies report on pollution as a marginal activity: they are on other business and may perform the function vis à vis oil pollution at low marginal cost. One or two countries do have specialized pollution inspectorates, and a number follow up reports of pollution at sea by sending out aircraft and helicopters to gather evidence. While it has been suggested that the liability provisions contained in the International Convention on Civil Liability for Oil Pollution Damage adopted in Brussels in November, 1969, might, when ratified, give rise to forms of private policing of pollution provisions by the insurance companies or tanker owners, we have seen that there is no sign of insurance companies at least instituting their own "watchdog" supercargoes.

To prevent pollution where the waste disposal is of an occasional or a once-and-for-all nature, many countries require official notification of the intention to deposit waste, and will then give instructions as to the types of containers to be used, the areas where the waste is to be disposed and so on. Such a control mechanism helps limit pollution offenses. It assumes particular importance where the detection of polluting offenders is so naturally difficult.

One area where the control of waste disposal on the High Seas has been internationally coordinated is that of the dumping of radioactive waste. Article 25 of the Geneva Convention on the High Seas requires states to prevent pollution resulting from radioactive waste disposal, and instructs them to cooperate with international organizations in doing so.[18] In May, 1966, the I.A.E.A. convened a panel of experts to discuss research and experience in radioactive waste disposal, and in 1967 five countries within the framework of the E.N.E.A. cooperated in an experimental operation to dump 10,893 tons of waste in the Eastern Atlantic, about 450 kilometers from the nearest land. The actual dumping was preceded by a hazard assessment of the dumping area by a group of experts, and the nuclear centers concerned subscribed to an insurance scheme to guard against damage arising from the operation with a ceiling of five million pounds. A similar operation was carried out in 1969 with not only officials from the waste disposing countries, but escorting officers from West Germany, Ireland, Japan, and Portugal. These developments may be seen as a form of auto-policing.

RAW MATERIAL EXPLORATION AND DEVELOPMENT

The search for and development of raw materials from the seabed is almost entirely undertaken in areas of national authority, even though this authority may not be internationally recognized.[19] The constabulary function is thus primarily concerned with policing rules, laws, and regulations which are exclusive to the zone of operations. For this marine activity at least, there are few of the problems of law enforcement on the High Seas which we have noted in other sections.

National codes for exploration and development have a double concern. First they aim to protect the operator's right of quiet enjoyment in the concession area. Second they seek to ensure the proper conduct, both technically and socially, of the concession by the operators. In respect to the first point, the protection of the operation from unnecessary nuisance from other marine users, there appear to be few specialized constabularies. In most countries it would be branches of the armed forces which would be called out for such protection duties during peacetime. There appears to be no direct parallel at sea to the occasion when British Petrol carried guns to protect themselves in Libya, though De Beers do operate security forces to enforce their rights over diamonds in the territorial waters of South Africa. Certainly as regards oil and natural gas operations, peace time threats to offshore quiet enjoyment are rare, and as a senior officer in one of the international major oil companies put it, where they do occur all that is needed is a fast boat with a submachine gun.[20]

The task of policing the second form of national codes is more substantial. In the case of oil and natural gas, these codes are usually embodied in the contracts signed by the operating companies with governments. They cover good oil field practice (in such matters as deviation drilling, unitization, the abandonment of boreholes and so on), the limiting of nuisance to other users of the sea (by properly lighting offshore installations, supervised use of explosives for seismic surveys, or strict safety measures to prevent pollution), the provision of good working and living conditions, the adoption of adequate safety arrangements, and the supply of full information to the government about the results of surveys, the progress of operations, etc.[21]

The oil companies themselves claim that there is no intrinsic enforcement problem with regard to these codes, since they have an interest in respecting them just as strong as have the governments. In some cases the point has substance, and is reflected in the existence of intra-company control systems, for example, the Shell Group Safety Committee, which are a form of auto-policing. But in other fields there is no such manifest coincidence of interests. The provision of full information to governments is one. Another was recently exemplified in the Gulf of Mexico disaster, caused by oil leaking from seven wells belonging to a subsidiary of Standard Oil, California. The Department of the Interior claimed that this would have been prevented had a safety valve required by regulations been in place, and the subsidiary acknowledged that 120 of its 292 offshore wells did not have these required valves.

In the Netherlands an inspector of fisheries acompanies each exploration vessel in order to enforce the limitations on the type of explosive that can be used in prospecting.[22] Survey vessels working off the shores of Honduras and Nicaragua are required to call in at a named port to pick up inspectors charged with monitoring the survey.

In general we may note a sharp contrast between the attention that has been given both nationally and internationally to the elaboration of codes for oil and gas exploration and development and the marked lack of discussion of enforcement procedures. It is notable in this respect that the sophisticated Pro-forma Regulations for the Conservation of Petroleum Resources drafted by the Organization of Petroleum Exporting Countries for adoption by its member countries leaves the enforcement of the provisions entirely up to the states themselves to plan out.[23] However, the rapid increase in offshore drilling, the growing awareness of the dangers of pollution from offshore operations, and the continuation of potential conflict between offshore petroleum operators and other users of the sea and the seabed, all suggest the importance of developing adequate enforcement procedures for offshore petroleum codes.

SUBMARINE CABLES AND PIPELINES

The articles of the Geneva Convention on the High Seas in respect to submarine cables and pipelines contain three main points: that there is a basic freedom to lay submarine cables and pipelines in international waters, that willful or culpably negligent damage to such lines of communication should be a punishable offense, and that owners of cables and pipelines should compensate owners of ships who have sacrificed gear in order to prevent injury to a particular cable or pipeline.[24]

The constabulary problem arises little in respect to the first of these points. There is as yet no significant conflict of interest which might lead to a challenge to the freedom to lay cables and pipelines in international waters, and within territorial waters it appears to be common practice to recognize this freedom subject to notification of the public authority. In Britain for instance, the Board of Trade must be notified of proposals for laying cables, and it will normally give consent after consulting other interested parties (Trinity House, Ministry of Agriculture and Fisheries, Commissioners for Crown Lands, and so on).

The main problem centers round the problem of damage to laid cables and pipelines. In spite of compensation provisions aimed at removing the motivation for an offense, and in spite of the provisions of the 1884 Convention for the Protection of Submarine Cables allowing public ships of other contracting states to require the exhibition of a ship's papers as evidence of nationality, and to draw up reports for presentation as evidence of alleged infringements, there is still heavy damage to submarine cables and little action against offenders.[25]

What constabulary forces there are for the enforcement of the high seas provisions are provided not by public bodies but by the owners of the cables themselves. Since the late 1950's they have operated a North Atlantic patrol for the policing of the Newfoundland fishing area. The patrol is conducted by a ship from the fleet of one of the cable owners, which is supplemented by chartered aircraft. Ships from the fleets of different cable owners will perform the function of the patrol ship according to an agreed rota. Their main functions are to warn fishermen who are fishing near the cables, gather evidence of infringements, and repair any damaged cables. The patrol is organized privately, though the twenty-one leading cable owners are formally linked through the International Cable Protection Committee. The task of this committee is to coordinate all measures for the protection of international communication cables against accidental interruption, including the charting of cables and the informing of fishermen, ocean scientists and other sea users about cable positions.

The submarine cable constabulary function is therefore privately performed. Its effectiveness is difficult to assess. There are a significant number of claims by fishermen operating the smaller less powerful vessels for compensation for loss of gear, but there is no statistical evidence to indicate how far these have been affected by the patrol and the work of the I.C.P.C. Certainly the number of prosecuted offenses for damage to cables is negligible: in a recent case a French trawler was convicted and fined ten thousand dollars for damage to a cable on the High Seas, but this is a rarity. Finally it should be remembered that while all cable ships operate an informal patrol while pursuing their other activities (there are fifty-six registered cable ships worldwide which are concerned variously with laying, maintaining and repairing cable) the only formal coordinated patrol is in the Newfoundland fishing area. Whether this will be extended depends partly on the continued growth rate and competitiveness of submarine telephone cables as against communication by satellite, partly on the development of other uses of the sea which might endanger the cables, and partly on the cost of patrolling in relation to these.

OCEAN DATA COLLECTION

There is no generally accepted international law on Ocean Data Acquisition Systems, though a draft Convention on their legal status is currently under discussion.[26] The absence of such has caused particular concern with respect to the willful damage and removal of O.D.A.S. As an I.O.C. Group paper put it: "[at the present time it is] highly undesirable to leave unmanned surface O.D.A.S. unattended in certain waters. Numerous cases have been reported in the last few years of interference with O.D.A.S. or equipment thereon . . . and it is clear that many of these cases amount to deliberate theft."[27] The problem is compounded in some countries where salvage is payable on the return to the owners of O.D.A.S., a provision which increases the incentive to larceny. Even where national laws do cover the protection of O.D.A.S. there is no specific system of inspection: the very difficulty of policing unmanned O.D.A.S. led the I.O.C. group of experts to suggest that the offense should be established and inspected at one remove by making it a criminal offense to be found in the possession of the whole or a part of an identifiable O.D.A.S. "in circumstances that suggest the commission of a criminal offense," if the person cannot prove lawful possession.[28]

BROADCASTING

There have recently been a number of cases of the extra-territorial sea being used as a base for broadcasting stations. Such operations were prohibited under provisions 422 and 962 of the Radio Regulations, Geneva 1959, and quasi-constabulary powers for the enforcement of these regulations were invested in the International Frequency Registration Board, a permanent organ of the International Telecommunications Union. The I.F.R.B. have no powers of enforcement per se. Rather, they investigate alleged contravention of the regulations, contact the Telecommunication Authorities of the country which had registered the ship in question, and informally encourage sanctions against evident offenders. The Telecommunications Administrations of the countries approached by the I.F.R.B. have all complied with the 1959 Provisions. Not only have none of them issued a license to a broadcasting transmitter on board a ship registered in their country, but on receipt of information that a ship registered in their country was carrying out illegal broadcasts, they have immediately revoked the wireless licenses which had been issued to the ship for normal ship communications. Further, they have taken action through the Maritime Authorities of their countries with a view to the cancellation of the registration of the ship itself.[29]

The 1959 Geneva Administrative Radio Conference also adopted a recommendation asking "Governments to study possible means, direct or indirect, to prevent or suspend such [extraterritorial marine] operations, and where appropriate, take the necessary action." [30] A number of countries have passed laws to this effect, prohibiting broadcasts from the open sea. The basic strategy of most of these laws is not only to prohibit the transmission of broadcasts, but to outlaw effectively the transmitters in the medieval sense of the term, i.e., to make it an offense to have any dealings with the transmitting offender. The 1962 Finnish law for example specifies such dealings. It makes liable to fine anyone who promotes an unlawful marine broadcast by financial support, by delivering, using, repairing or maintaining technical installations or other objects for this purpose, by supplying or procuring materials for broadcasting, by providing transport to the ship on which the installation is situated, and by taking part in a broadcast on board. Similar provisions are contained in the Swedish, French, Belgian, Danish, and British laws on the subject.

INFRINGEMENTS OF NATIONAL BOUNDARY LEGISLATION

The use of the sea in the course of the smuggling of goods, the evasion of exchange controls, or the bypassing of migration regulations is merely as an area of transit. Since the purpose of these offenses is to transfer goods, money or people from one land area to another, many of the constabulary functions may be performed on land, either in the country of departure or in that of destination. Since the evasion of any nation's controls on incoming traffic will not yet have been committed in the country of departure, the *de facto* inspectorates there will be in the nature of early warners (in the case of overseas informers on intended smuggling for example) or of preventers (as in the case of the checking of passengers' landing credentials by shipping companies and airlines, who stand to be penalized by a fine, by the cost of keep, and by the marginal cost of the return journey, for any passenger landed in the U.S.A. without a satisfactory visa).[31] Within the country of destination, inspection, excluding coastal inspection, will take the form of checking import permits or duty receipts on suspected goods, or checking identity cards, labor permits, hotel registers, passports and so on, for immigrants.

We have suggested that coastal policing should be seen as only one part of a more general policing system, whose importance will vary with the feasibility of a coastal-barrier strategy, and also with the political acceptability of intra-boundary controls. Despite these variations, most maritime countries have significant coastal patrols. In general they follow the principle of channelling incoming goods, money, and persons to specialized inspection points and operating patrols to ensure that these channels are observed.

Few of the specialized inspectors are seaborne. Customs officials travelled on the *Queen Mary* and *Queen Elizabeth* in post-war years up to 1952 in order to speed up clearance, but this was considered uneconomic and was stopped. Immigration officials continued to travel on the trans-Atlantic routes until 1969 and still do travel on the routes across the North Sea, again to speed clearance. In small-boat and merchant shipping, customs checks will be usually made on board, and most ports will have customs launches for this purpose. However, the great proportion of specialized inspection posts will be on land or at the dockside.

The inter-inspection-post patrols on the other hand often involve considerable seagoing activity. The British Water Guard, a division of H.M. Customs and Excise, has three revenue cruisers, with a usual manning of six per vessel, plus motor launches for work in creeks, and the above-mentioned boarding launches. In the U.S.A. the Coast Guard have an extensive fleet of cutters, which work in cooperation with the U.S. Immigration Service, and have a right to board any U.S. vessel on the High Seas which is believed to be violating or has violated U.S. laws. The French Customs have a developed coastal patrol which uses helicopters as well as boats, and in Hong Kong, where the comparative wealth and the tight restrictions cause particular pressure on national boundary legislation, extensive patrols are operated by the water guard, the immigration service and the police.[32]

Sea-based patrols are also used in the policing of controls on the outflow of goods, money, and people from a country. Soviet naval vessels on general patrol in the Baltic, watch for vessels with suspectedly unauthorized emigrants on board, and coastal watches are similarly kept from other countries with emigration and exchange control restrictions.

In general, while the seaborne patrols form only a part of the constabulary system for the enforcement of national boundary legislation,[33] they may nevertheless in some countries be the largest seaborne inspectorate of all those we have up to now discussed. Certainly this is true of Hong Kong, as of Abu Dhabi whose national navy is concerned primarily with the enforcement of customs and migration controls. Changes in technology and factor costs may cause a shift away from seaborne to land-based patrols, as has happened in Britain. But as long as the maintenance of a country's economic wealth is seen as particularly dependent on the effective enforcement of boundary controls (as in the cases of Hong Kong or Abu Dhabi cited above or East Germany as regards labor or indeed most underdeveloped countries with accumulated capital as regards exchange control) then we may expect sea patrols of the kind we have discussed to continue as far as maritime states are concerned.

CONCLUSION

The foregoing discussion has two concerns: first, it has sought to clarify whether the policing of sea and seabed activities is in any substantial way distinct from the policing of society on land; and second, it has enumerated the types of constabularies and inspectorates which are concerned with nonmilitary uses of the sea and seabed whose existence and practice should be noted in the course of designing verification measures for marine arms control.

With respect to the first of these points, our discussion suggests that there are features which distinguish the policing of marine activities from other forms of policing. First, the fact that the sea is an appendage to land and that all users of the sea return to land, means that many of the constabulary functions both of detection and arrest can be carried out on land. This presupposes that detection takes place by imputation (as in the case of the inspection of fishing gear, fish sizes, oil tanks, or ship's equipment) or that the offense is committed on landing and may therefore be detected on land (as in the case of smuggling, illegal immigration or, in a slightly different form, pirate radio broadcasting). It also presupposes that any offender returns to a land area where the political authority has an interest in arresting him. These presuppositions commonly, but by no means always, hold.

Where detection and arrest cannot properly be carried out from a specified land area, a second distinguishing feature of the seas becomes relevant, namely the freedom of the High Seas. As far as detection is concerned, much information about offenses can be obtained by observation without infringing the freedom of the High Seas. In some cases, however, on-board inspection may be desirable and this does raise the issues of freedom of the seas and sovereignty where it is a question of inspecting the vessels of other flag states. Some of the Fishery Conventions have made provisions for "cooperative" inspection systems because of the high cost of "atomistic" policing, but the restricted range of such "cooperative" powers and the failure to develop "integrated" inspectorates underlines the difficulties of developing effective enforcement procedures in zones of restricted jurisdiction.

Thirdly, while users of the sea must necessarily return to land, the fact that high seas are international waters, i.e., not enclosed within a single political authority, means that transgressors of marine laws may be free to return to a land area whose political authority may have no powers nor interest in enforcing the laws in question. The constabulary problems arising from this are reflected in the experience of pirate radio broadcasts and fishing limit offenses by foreign craft. Effective enforcement in these cases depends on common agreement between the states bounding the sea to seize and sanction vessels seeking to escape from particular national jurisdictions.

Fourthly, natural, technological and economic factors have combined to make the average cost of general surveillance higher for marine activities than for those on land. This has led to a further pressure to devise means whereby marine activities can be policed on shore, and where this cannot be done, reliance has been placed on low-cost forms of information such as that deriving from other users of the sea or air-space, both public and private, who are instructed, encouraged or paid for the reporting of suspected transgressions.

Historical, and even contemporary parallels to these four features can be found for land areas, but taken together they do constitute a distinct set of conditions of which account must be taken in the elaboration of enforcement measures for marine activities.

Turning to the second purpose of our paper, we have found that there are already in existence a substantial number of constabularies and quasi-constabularies concerned with the enforcement of marine laws and regulations. The points we have found common to most national systems are: (1) a notable absence of consolidation; (2) the predominance of land-based inspection, for reasons discussed above; (3) the rarity of on-board inspectorates traveling with the vessel; (4) the paucity of "cooperative" or "integrated" constabularies in the sense defined above.

One final point should be made. We have dealt with the constabulary function as composed of two parts:

(a) the obtaining of information about transgressions (this is a more general formulation than "boarding and inspection").

(b) the arrest and escort of suspected transgressors to zones of penal jurisdiction.

These are, however, only two aspects of any enforcement system, and it is important when considering the procedures to adopt and the resources to allocate in respect to these two functions to see them in the context of all the factors relevant to enforcement. We may list these other factors as follows:

(c) making the law known (pollution regulations are included in exams for mariners in many countries, so is submarine cable law and the positioning of cables; informatory leaflets are issued in respect to immigration regulations, pollution, customs controls and so on).

(d) removing the incentive to break the law (this is the point of submarine cables compensation, or the provision of port facilities for the disposal of waste oil).

(e) encouraging the desire to obey the law quite apart from fear of detection and penalization (this is a central aspect of the literature on organizational control, and is one aim of policies encouraging participation in rule-making by those to whom the rules will apply).

(f) instituting technological changes which make it difficult or impossible to disobey the law (this occurs with the regulated net size in fisheries, the limitation of the size of fishing vessels, the sinking of submarine pipelines and cables underground, the design of whalers so that they can only process whales of a legitimate type, or the design of tankers to make intended oil pollution impossible).

(g) The conviction of the offender.

(h) the punishment of the offender, including the size and type of punishment, and the person or persons on whom the punishment will fall.

(i) the mitigation of the effects of the offense (as in the counteracting measures used against pollution, the emergency regulations governing conduct and help after shipping accidents, or the jamming of pirate radio broadcasts).

For each of these there will be some optimal method of operation. Thus if the obtaining of information involved (a) sea patrols, (b) air patrols, and (c) onshore inspectors, there will be, for a given probability of detection, some optimal mix between the three which will minimize the cost of inspecting. Similarly for a given level of enforcement, there will be an optimal mix between the enforcement factors which will minimize the cost of enforcement, and an optimal level of enforcement given the minimized costs of enforcement and the benefits deriving from it. What is noticeable in many marine activities, above all in fishing and pollution, is that enforcement procedures have not been devised with a full awareness of the economies to be gained from such an optimizing procedure. Certainly there are problems and costs of obtaining information for implementing this approach, but the real point is that the consciousness of the principle that the costs and the benefits of enforcement procedures should be equal at the margin, and that there is some degree of substitutability between the different types of enforcement procedure, will prompt the designer of an enforcement system to ask questions which he might otherwise not have asked. On the evidence of the existing constabularies and inspectorates which we have discussed, and in the light of the current proposals for verification measures for marine arms control agreements, it is a point which it is difficult to over-emphasize.

NATIONAL CLAIMS TO TERRITORIAL SEA AND FISHING LIMITS (1969)

Country	*Territorial Sea*	*Fishing Limit*	*Remarks*
	(Nautical miles unless otherwise stated)		
Albania	10	12	From straight base-lines.
Algeria	12	12	
Argentina	200	200	Permission must be sought by boats of foreign flags to carry out fishing activities at a distance of not less than twelve miles from the coast.
Australia	3	12	
Belgium	3	3	Customs control zone ten nautical miles.
Brazil	12	12	
Bulgaria	12	—	
Burma	12	12	From straight base-lines.
Cambodia	12	12	From straight base-lines.
Cameroon	18	—	
Canada	3	12	
Ceylon	6	106	From "appropriate base-lines." Fishing claim is to conservation zones.
Chile	3	200	Contiguous zone twelve nautical miles.
China (Nationalist)	3	—	
People's Republic of China	12	—	From straight base-lines.
Colombia	3	12	Contiguous zone twenty kilometers.
Congo (Brazzaville)	3	—	
Congo (Kinshasha)	3	—	
Costa Rica	200	200	

Country	*Territorial Sea*	*Fishing Limit*	*Remarks*
	(Nautical miles unless otherwise stated)		
Cuba	3	–	
Cyprus	12	12	
Dahomey	12	12	
Denmark	3	12	From straight base-lines. Fishing limit three nautical miles. South of Kattegat.
Dominican Republic	6	12	Contiguous zone twelve nautical miles.
Equatorial Guinea	–	–	Not known. Six nautical miles can be assumed.
Ecuador	200	200	
El Salvador	200	200	
Ethiopia	12	–	Straight base-lines around Dahlak Archipelago.
Faroes	3	12	Fishing limit from straight base-lines.
Finland	4	–	Employs straight base-lines.
France	3	12	Straight base-lines are employed around Brittany and parts of the Mediterranean coast and around the west and south coasts of Corsica.
Gabon	12	–	
Gambia	3	–	
Germany, East (not recognized by HMG)	–	–	Not known but three nautical miles can be assumed from straight base-lines.
Germany, Federal Republic	3	3	Contiguous zone ten nautical miles.
Ghana	12	12	May claim certain areas out to 112 nautical miles as Fish Conservation Zones.
Greece	6	6	Fishing reciprocity between three and six nautical miles.
Greenland	3	12	From straight base-lines. Outer six nautical miles of fishing limit phased out until May 31, 1973.
Guatemala	12	–	
Guinea	130	130	
Guyana	3	–	
Haiti	6	–	
Honduras	12	12	
Iceland	3 or 4	12	Breadth of territorial sea uncertain. Fishing limit from straight base-lines.
India	12	112	Fishing claim is to conservation zones.
Indonesia	12	–	From straight base-lines enclosing whole archipelago.

Country	*Territorial Sea*	*Fishing Limit*	*Remarks*
	(Nautical miles unless otherwise stated)		
Iran	12	–	From straight base-lines twelve nautical miles apart.
Iraq	12	–	
Irish Republic	3	12	From straight base-lines.
Israel	6	6	
Italy	6	6	
Ivory Coast	6	12	Contiguous zone twenty kilometers.
Jamaica	12	–	
Japan	3	–	
Jordan	3	–	
Kenya	12	12	From straight base-lines.
Korea, North	12	–	Probably employs straight base-lines.
Korea, South	–	20-200	Territorial sea not known but three nautical miles can be assumed probably from straight base-lines. Fishing limit now believed to be twelve nautical miles.
Kuwait	12	–	
Lebanon	6	6	Customs control zone twenty kilometers.
Liberia	12	–	Contiguous zone twenty-four nautical miles.
Libya	12	–	
Malagasy Republic	12	–	From straight base-lines.
Malaysia	12	12	
Maldives	3	–	Measured from rectangle enclosing whole group.
Malta	3	–	
Mauritania	12	12	From straight base-lines.
Mauritius	3	–	
Mexico	9	12	
Monaco	3	–	
Morocco	12	12	Claims only six nautical miles fishing limit in Straits of Gibraltar. Employs bay closing lines limited to thirteen nautical miles.
Muscat and Oman	3	–	
Netherlands	3	–	
Surinam	3	–	
New Zealand	3	12	
Nicaragua	–	200	Reported to claim three nautical miles territorial sea.
Nigeria	12	12	
Norway	4	12	From straight base-lines. Outer six nautical miles of fishing limit phased out until October 31, 1970. Customs control zone ten nautical miles.

Country	*Territorial Sea*	*Fishing Limit*	*Remarks*
	(Nautical miles unless ohtrewise stated)		
Jan Mayen	4	—	From straight base-lines.
Pakistan	12	—	A fish conservation zone extending one hundred nautical miles beyond territorial sea is claimed.
Panama	200	200	
Peru	200	200	
Philippines	Special	—	Limit of territorial sea is the limit set out in the Treaty of Paris 1898. Straight base-lines enclose whole archipelago for internal water purposes. Fishing limits the same as territorial sea limits.
Poland	3	—	
Portugal	6	12	Overseas territories follow same law as mother country: Angola, Mozambique, Timor, Portuguese Guinea, Cape Verde Islands, Sao Tome, and Principe Islands, Macao.
Republic of South Africa	6	12	
Romania	12	—	
Saudi Arabia	12	—	From straight base-lines, to islands twelve nautical miles apart.
Senegal	12	18	From straight base-lines. Outer six miles of fishing limit not enforced against states conforming 1958 Law of the Sea Conventions.
Sierra Leone	12	12	
Singapore	3	—	
Somalia	12	—	
South Yemen, People's Republic	—	—	Decree to be issued shortly.
Spain	12*	12	*For fiscal purposes. Limits apply to all Spanish territories including enclaves on Moroccan coast.
Sudan	12	—	
Sweden	4	4	From straight base-lines. Fishing limit two nautical miles on west coast.
Syria	12	—	
Tanzania	12	—	
Thailand	12	12	
Togo	12	12	
Trinidad and Tobago	3	—	
Tunisia	6	—	Fishing limit is partly twelve nautical miles and partly out to fifty meter depth contour.
Turkey	6	12	From straight base-lines.

Country	*Territorial Sea*	*Fishing Limit*	*Remarks*
	(Nautical miles unless otherwise stated)		
United Arab Republic	12	12	From straight base-lines, to island twelve nautical miles apart. Claims further six nautical miles contiguous zone beyond fishing limit.
United Kingdom	3*	12	*Including self-governing colonies and protectorates whose foreign affairs are the responsibility of the U.K.
Bahrain	3	—	Twelve nautical miles fishing limit under consideration.
Brunei	3	—	
Qatar	3	—	
Tonga	3	—	
Trucial States	3	—	(Abu Dhabi, Ajman, Fujaira, Ras Al Khaima, Sharja, Umm al Qaiwain)
Uruguay	12	*	*Fishing limit extends to edge of Continental Shelf.
U.S.A.	3	12	
U.S.S.R.	12	—	From straight base-lines.
Venezuela	12	—	From straight base-lines. Reported to claim fifteen nautical miles fishing limit.
Vietnam, North	—	—	Not known but twelve nautical miles can be assumed.
Vietnam, South	3	20 kilometers	Contiguous zone twelve nautical miles.
Western Samoa	3	—	
Yemen	—	—	Not known but twelve nautical miles can be assumed.
Yugoslavia	10	19	From straight base-lines. (Claims two nautical miles contiguous zone beyond territorial sea).

ALVA MYRDAL

22 No Arms on the Ocean Floor

The activities of the Geneva Disarmament Committee—the CCD—are most crucial today as it has the task of establishing rules guaranteeing that the seabed under international waters be excluded from the arms race. It is of course, a sad comment on the mad state of world affairs that so much labor has to go into efforts to prevent death machines from being installed in the great ocean depths, because this, nothing less, but also nothing more, is what the CCD is actually at work on: an international agreement on, as the text goes, the prohibition of the emplacement of nuclear weapons or other weapons of mass destruction on the seabed and the ocean floor and in the subsoil thereof.

Thus, what engages us is, strictly speaking, an attempt to legislate a non-armament measure, that is, to forestall a military development which has not yet occurred. Further, to do so only in regard to ABC weapons (atomic, biological, and chemical) and, finally, to prevent only their emplanting or emplacement. This restrictive framework for the negotiations in their present phase has given rise to some comments, criticism, but also constructive suggestions, and some of these I wish to dwell on with somewhat more specific detail.

But first, one more prefatory note of a general character. The

---

Alva Myrdal is a Swedish cabinet minister, diplomat, sociologist and writer. Among her many positions, she is Minister for Disarmament, Minister for Cultural Affairs, and a delegate to the Conference of the Committee on Disarmament (CCD). She is the wife of Karl Gunnar Myrdal, the noted sociologist.

material possibility of engaging in an armaments race on the ocean floor rests really with two nations only, the super-powers—the Soviet Union and the United States—as they are the only ones who possess today the technical and economic capabilities of reaching that deep ocean floor. For the purpose of what is presently being negotiated, a bilateral agreement between these two would have sufficed. This judgment could be based on an argument which is often proffered by the spokesmen of these two powers against requests from some other states that the scope of the prospective prohibitions should be widened. They say, always: with the present state of the art, anything more would be premature.

Well then, what interest can other countries have in this issue? Why are in Geneva twenty-three other delegations taking such a lively part? First, because it does no harm and may even be worthwhile for the future to have more adherents to a treaty even if it cannot materially concern them very much today. I then disregard the fact that some sea mounts and sea ridges might be accessible also to some other states. But, secondly, and much more important, we are interested because all nations of the world must be assured that a promise of non-armament, just as any promise of disarmament, is kept.

We cannot afford to be fooled. Thus, the need for control, for verification, is the main reason for our interest. To be able to share in the surveillance of international obligations is a very strong motive for most states to become parties to this kind of treaty. And then, the overwhelming reason why all states must take an interest in this particular treaty is that it would be the first one to launch the supreme principle that no usurpation by nations or other interest groups be allowed in regard to any area of the deep ocean floor. Thus, as a matter of fact, the disarmament committee is busy at work to pave the way for what is the task of the U. N. Seabed Committee, that is, to establish an international regime for the world's last frontier province.

I now wish to give a short description of the Draft Treaty as it is worded in the version presently before the CCD. This text, which was introduced on April 23, 1970, by the Delegations of the

Soviet Union and the United States jointly, represents the second attempt on their part to find language which could be generally accepted by other member-states. It is probably not the last attempt. A more final version of the treaty is generally expected to appear in order that the treaty should be hopefully unanimously presented by the CCD to the General Assembly of the United Nations in autumn, 1970, for endorsement by all member-states.

The general aim of the treaty is expressed in the preamble. It states that the treaty shall facilitate the peaceful exploration and use of the seabed and the ocean floor in the common interest of mankind. At the same time, it shall prevent the seabed and the ocean floor from being drawn into the nuclear arms race. The states party to the treaty then undertake not to emplant or emplace on the seabed and the ocean floor and in the subsoil thereof any nuclear weapons or any other types of weapons of mass destruction. Even structures, launching installations, and other facilities specially designed for storing or testing or using such weapons are prohibited.

The prohibitive rules are valid for the whole of the ocean floor and the seabed beyond a twelve mile coastal zone and also within that zone for all parties except the coastal state. In this so-called seabed zone the coastal state would have all freedom of action even to emplace weapons of mass destruction. That is of course a prerogative of importance only to those five nations who possess nuclear weapons.

I am not going to enumerate the parts of the treaty text as they are hardly of substantial importance. I just find one article worth special mention, and that is Article 6 which establishes a pattern of review conferences. Five years after the entry into force of the Treaty and repeatedly in the future if so requested, a conference shall be held with the view of assuring that the purposes both of the preamble and the provisions in the Treaty are being realized. Thus, the review conference constitutes a kind of safety valve.

So far I dealt with Draft Treaty text as it actually is before the CCD. I should now like to turn to some points of criticism and suggestions for changes. I can, of course, not refrain from reflecting more the attitude of the smaller and particularly the non-aligned nations. It is the non-aligned, but not solely these, who are still upholding some requests as to further amendments of the text, and I wish to mention just a few of those points because they concern questions of principles and future reforms which will undoubtedly be discussed during this convocation. One of the points relates to the fact that the Treaty in effect leads only to denuclearization of the seabed, not to demilitarization which has all along been the goal of the vast majority of states. This is the main reason why many delegations have come to attach such importance to the insertion in the treaty of a pledge obliging the parties to continue negotiations toward reaching further prohibitions in the seabed. Even if the so-called crawlers are intended to be covered by the present prohibitions, there are in the offing many new developments in regard particularly to anti-submarine warfare, including even mines which seem to many of us as menacing the main goal of insuring peace in the oceans and preserving them as the common heritage of mankind.

The Swedish Delegation suggested last autumn both in the CCD and in the United Nations a somewhat more far-reaching formula than the one contained in the preamble by way of a new article in which the parties would pledge themselves "to continue negotiations in good faith on further measures relating to more comprehensive prohibition of the use for military purposes of the seabed and the ocan floor and the subsoil thereof." That wording was largely based on the similar provision regarding further negotiations to curb the nuclear arms race which was inserted in the text of the Non-Proliferation Treaty during the negotiations on that subject, on the insistence of the non-nuclear-weapons states. Several delegations have been asking for the inclusion of such a formula in the final text. One of the reasons is, of course, the fate which a similar pledge being placed just in a preamble to an earlier arms control measure has had. I am referring to the paragraph in the preamble of the Partial Test-Ban Treaty in which the parties did express their determination "to achieve the discontinuance of all test explosions of nuclear weapons for all time, determined to continue negotiations to this end." That was a pledge in the preamble, but we all know that during the seven years elapsed it has not been fulfilled and that the actual course of developments goes in the direction instead of increases particularly in the field of nuclear tests. So disappointment has taught us a lesson.

Another point of crucial concern to all of us when we want to plan for a stronger U.N. influence in the world affairs of the future, concerns the desire to internationalize the control. From the outset of the negotiations in the CCD, a link has therefore been suggested with the envisaged future international machinery which would have to ensure that the exploration and exploitation of the natural resources of the seabed would be carried out in a way which furthers the implementation of the principle of joint benefits for all nations. The same machinery could safeguard the verification provisions of a disarmament treaty. Already last spring the non-aligned members proposed that when it became feasible, verification could be carried out not only by the individual parties, but also through an appropriate international agency or arrangement.

As we interpret this suggested provision, it was a way of saying, although admittedly very indirectly, that if and when an international machinery for the seabed was set up it might be possible for states so desiring to make use of that machinery for its verification needs in relation to the treaty we are now discussing. In the new draft text, which in other respects closely reflects the contents of the Canadian proposals, these references to international good offices, including those of the Secretary-General, had been omitted.

In view of the importance being attached to the principle of the seabed representing a common heritage of mankind, and the link between that principle and the notion of an international regime for the seabed, many delegations regard the failure to mention at all in the new text the possibility of international controls as a serious weakness. They continue to appeal for the incorporation into the final text of some reference to the idea of international verification as a possible future development. One delegation of Eastern Europe has further suggested that the CCD should keep on its agenda the question of the demilitarization of the seabed and the ocean floor. In that way members of the committee could raise the question of further steps; perhaps a further treaty leading to this end whenever they see that the question is ripe for discussion without waiting for the review conference which is provided for in the treaty.

While the present phase of the work in the Geneva Disarmament Committee is restricted to an attempt to safeguard denuclearization of the seabed, there is a strong feeling among a majority of delegations that negotiations must proceed to a total ban even if a treaty on a partial ban is first finalized. National militarization and international peaceful exploitation of the riches of the seabed simply cannot coexist. The overriding aim which I believe has already begun to echo through this conference is to secure the seabed as the common heritage of mankind and for this as well as for keeping the seabed free from an armaments race and free from international conflicts in general, an international regime with some international administrative machinery is necessary. Only so can we rest assured that the further exploration and exploitation of the natural resources of this very important area will be undertaken in a way to further the interests of all states, and these interests flow together in the quest for peace in the oceans and the vision of wealth in the oceans.

APPENDIX:

THE OCEAN REGIME

PROPOSAL

ELISABETH MANN BORGESE

THE OCEAN REGIME
DRAFT STATUTE
(Revised, December 1970)

ARTICLE I
ESTABLISHMENT OF THE REGIME

The Parties hereto establish an International Regime for the Peaceful Uses of Ocean Space (hereinafter referred to as "the Regime") upon the terms and conditions hereinafter set forth.

ARTICLE II
DEFINITIONS

1. *Ocean Space* shall include the high seas, the territorial waters and contiguous zones; the atmosphere above it; the Continental Shelf; the seabed and what is below it.
2. *International boundaries* are boundaries between nations, protecting the territorial integrity and national sovereignty of any one nation against intrusions or exploitations by another nation or other nations or their citizens. There are no boundaries between a nation and the international organization or community of which that nation is a part.
3. *The High Seas* shall include international waters beyond the limits of territorial seas as defined by common accord by a duly constituted international Conference.
4. *The seabed* and what is below it shall extend to the outer limit of the legal continental shelf.

ELISABETH MANN BORGESE is a Senior Fellow at the Center for the Study of Democratic Institutions, Santa Barbara, California, and Chairman of the Planning Council of the International Ocean Institute, Malta.

5. *The legal Continental Shelf* shall be delimited by the same line delimiting the territorial sea.
6. *The natural resources* referred to in this Statute shall include minerals, metals, and other nonliving resources of the seabed and below the seabed as well as living resources of ocean space, both animal and vegetal.

ARTICLE III
FUNDAMENTAL PRINCIPLES

A.

1. Ocean space is an indivisible whole. Geological structures extend, currents and waves move, species migrate across the high seas and the ocean floor regardless of political boundaries. The law of the seas and the seabeds must accord with this reality.
2. The High Seas beyond the limits of territorial waters and the seabed beyond the limits of the legal continental shelf as defined in this Statute are the common heritage of mankind. They are not subject to national appropriation by claim of sovereignty, by means of use or occupation or by any other means.
3. Activities of Nations or their subjects within the limits of the territorial seas or of the legal continental shelf, which might affect the ecology of ocean space or the common interest of mankind shall be subject to international regulation.
4. All states members of the Regime shall be subject to the International Mining Regulations and liable for any and all damages resulting from a violation of these Regulations or of any relevant rule of conventional or customary international law.
5. The natural resources of the High Sea and on or below the seabed as defined by this Statute are the common heritage of mankind. They must be developed, administered, conserved, and distributed on the basis of international cooperation and for the benefit of all mankind.
6. The use, exploration, and exploitation of the seabed beyond the limits of national jurisdiction shall be for peaceful purposes only. Nations shall continue in their efforts to validate this principle by extending it from the seabed to the superjacent waters, and from the international zone to the zones under national jurisdiction.
7. There shall be freedom of scientific investigation in ocean space. In zones under the jurisdiction of a coastal state such freedom shall be subject to the following conditions:

(a) that the authorities of the coastal nation must be informed sixty days in advance of the planned exploration;
(b) that experts of the coastal nation be invited to participate in the exploration;
(c) that any vessel used for the exploration, or any part of such vessel, must be open and accessible at all times to the experts from the coastal nation;
(d) that all data and all samples taken must be accessible to the experts from the coastal nation;
(e) that adequate measures must be taken to protect the coastal zone against dangers arising from seismic explosions, oil spills from deep drillings, or other hazards.

8. The use, exploration and exploitation of the seas and seabeds shall conform to international law, to the principles of the Charter of the United Nations, and to the Declaration on Principles of International Law Concerning Friendly Relations and Cooperation Among States in Accordance with the Charter of the United Nations, adopted by the General Assembly on 24 October, 1970.

B.

9. Member states shall bear international responsibility for national activities in ocean space, whether carried out by governmental agencies or nongovernmental entities, and for assuring that national activities are carred out in conformity with the procedures set forth in this Statute.
10. The activities of nongovernmental entities in ocean space shall require authorization and continuing supervision by the appropriate Member State.
11. When activities are carried out in ocean space by an intergovernmental organization or by an international or multinational organization or corporation ("Associate Member") responsibility for compliance with the provisions of this Statute shall be borne by such organizations themselves.
12. In the exploration of ocean space and the exploitation of its resources, Member States and Associate Members shall be guided by the principle of cooperation and mutual assistance and shall conduct all their activities in ocean space with due regard for the corresponding interests of all other Member States and Associate Members.
13. Member States and Associate Members shall render all possible

assistance to any person, vessel, vehicle, or facility found in ocean space in danger of being lost or otherwise in distress.

14. Member States and Associate Members engaged in activities of exploration or exploitation in ocean space shall immediately inform the Maritime Secretariat of any phenomenon they discover in ocean space that could constitute a danger to the life or health of persons exploring or working in ocean space.
15. All states shall have the right for their nationals to engage in fishing, aquaculture, in-solution mining, transportation, and telecommunication on and under the High Seas.
16. The rights declared in the preceding paragraph shall be subject to the treaty obligations of each Member and to the interests and rights of coastal states and shall be conditioned upon compliance with the rules established under Articles V,A,4 and VIII,E,7 of this Statute.

C.

17. The International Regime for the Peaceful Uses of Ocean Space shall provide a pattern for the future framework of international organization.

ARTICLE IV
OBJECTIVES

1. The Regime shall safeguard the ocean environment as an essential reservoir of life and shall transmit this common heritage of mankind legally intact and ecologically viable to future generations.
2. The Regime shall seek to harmonize the activities of science, industry and politics in the use of ocean space, and to this end:
   (a) it shall develop and enhance research and exploration of ocean space, and the contribution of ocean resources to the world economy;
   (b) it shall coordinate the activities and plans of all United Nations Special Agencies and other intergovernmental and nongovernmental international organizations engaged in the exploration and exploitation of ocean space and resources.
3. The Regime shall seek to harmonize the interests of all nations, regardless of their ideology or state of development, by increasing the participation of all people in the management of the ocean environment and its resources, and to this end:

(a) it shall take appropriate measures to protect developing Nations against the danger that might arise from a sudden drop of prices of minerals and metals consequent on progress in ocean-space technology;

(b) it shall take appropriate measures for the international training of experts, scientists, and technicians from developing Nations.

4. The Regime shall see that conditions are maintained that will encourage enterprises to expand and increase their ability to produce and to promote a policy of rational development of ocean resources avoiding inconsiderate exhaustion of such resources or pollution of ocean space.
5. It shall promote the improvement of the living and working conditions of the labor force in each of the industries under its jurisdiction.
6. It shall further the development of international trade.
7. It shall promote the regular expansion and modernization of production as well as the improvement of the quality, under conditions that preclude any protection against competing industries except where justified by illegitimate activities on the part of such industries in their favor.
8. It shall promote the development and harmonization of maritime law and international law relating to ocean space.

ARTICLE V
FUNCTIONS

A.

The Regime's institutions are authorized:

1. to regulate, supervise and control all activities on the high seas and on or under the seabed;
2. to accommodate conflicting uses of ocean space by setting priorities;
3. to fix shipping lanes and make other rules for navigation whenever the multiple use of ocean space so requires and otherwise to protect the freedom of navigation;
4. to make rules for the laying of submarine cables in order to avoid interference with mining or fishing operations or other uses of ocean space and otherwise to protect the freedom of laying submarine cables;
5. to determine universally applicable criteria for fisheries, fish farming, and aquaculture and, in cooperation with regional commissions

and organizations where such exist, to identify permissible fishing areas, establish fishing seasons, identify methods of capture, fix quotas, and specify types of resources that may be captured;

6. to issue licenses to Member States and Associate Members for the peaceful and orderly exploration and exploitation of the seabed and below the seabed, according to rules to be promulgated by the Regime;
7. to disseminate scientific information and to facilitate the transfer of technologies;
8. to review and revise the International Mining Code whenever changes in the ocean technologies so demand, and in particular
   (a) to issue regulations concerning pollution, waste of mineral resources and the disposal of radioactive waste materials in ocean space;
   (b) to promulgate safety standards for the exploration and exploitation of ocean resources, such as the use and equipment of drilling installations, the use of electrical installations, the use of radioactive equipment, the storage and use of explosives and the prohibition of detonating them in the vicinity of vessels engaged in fishing or in the vicinity of drifting or stationary gear or if schools of fish are discovered under or near the shot point; the equipment of vessels with radar, echo-sounder, and sonar, where seismic surveys are concerned; the prevention of fire, and the protection of historical and archeological discoveries;
   (c) to advise Member States with regard to safety regulations within their national jurisdiction so as to harmonize such regulations with those enacted for the area beyond national jurisdiction;
9. to inspect all stations, installations, equipment, and sea vehicles, machines, and capsules on or under the seabed;
10. to order license holders to suspend, modify, or prohibit activities or experiments if they might cause potentially harmful interference with the peaceful exploration and exploitation of ocean space;
11. to impose fines and cancel licenses if a Party violates the provisions of this Statute;
12. to propose development plans;
13. to impose an Ocean Development Tax;
14. to make its own budget, providing for its own administration and all other legitimate expenses; to accept loans and to make grants;

15. to establish Maritime Corporations as subsidiary organs, in accordance with Article XII;
16. to settle disputes between Member States, or between Member States and Associate Members; or between Member States or Associate Members and individuals; or between Member States or Associate Members or individuals, and the Regime; and to make awards;
17. to control inspectorates, constabularies, and armed forces operating on the seabed in accordance with Article XVI, in order to promote the objectives and ensure the observation of the provisions set forth in this Statute.

B.

In carrying out its functions, the Regime shall:

1. conduct its activities in accordance with the purposes and principles of the United Nations to promote peace and international cooperation and in conformity with policies of the United Nations furthering the establishment of safeguarded worldwide disarmament and in conformity with any international agreement entered into pursuant to such policies;
2. render decisions, recommendations, and opinions affecting the area directly subject to the jurisdiction of the Regime;
3. render recommendations and opinions affecting areas under the jurisdiction of coastal states or island states or affecting the atmospere;
4. allocate its financial resources in such a manner as to secure efficient utilization and the greatest possible general benefit in all areas of the world, bearing in mind the special needs of the underdeveloped areas of the world;
5. submit reports on its activities annually to the General Assembly of the United Nations and, when appropriate, to the Security Council. If, in connection with the activities of the Regime, there should arise questions that are within the competence of the Security Council, the Regime shall notify the Security Council, as the organ bearing the main responsibility for the maintenance of international peace and security, without prejudice to any measures the Regime is entitled to take under this Statute;
6. submit reports to the Economic and Social Council and other organs of the United Nations on matters within the competence of these organs.

C.

In carrying out its functions, the Regime shall not make assistance to Members or Associate Members subject to any political, economic, military, or other condition incompatible with the provisions of this Statute.

D.

Subject to the provisions of this Statute and to the terms of agreement concluded between a State or a group of States and the Regime which shall be in accordance with the provision of the Statute, the activities of the Regime shall be carried out with due observance of the sovereign rights of States.

ARTICLE VI
LEGAL STATUS

1. The Regime shall have juridical personality.
2. In its international relationships, the Regime shall enjoy the juridical capacity necessary to the exercise of its functions and the attainment of its ends.
3. In each of the Member States, the Regime shall enjoy the most extensive juridical capacity that is recognized for legal persons of the nationality of the country in question. Specifically, it may acquire and transfer property, and may sue and be sued in its own name.
4. The Regime shall be represented by its institutions, each one of them acting within the framework of its own powers and responsibilities.

ARTICLE VII
MEMBERSHIP

1. Members shall be States that deposit an instrument of acceptance of this Statute. Members shall be entitled to representation in the Maritime Commission, the Maritime Assembly, and the Maritime Court.
2. Associate Members shall be intergovernmental organizations or nongovernmental international organizations and corporations holding licenses issued by the Regime. Associate Members shall be en-

titled to representation in the Maritime Assembly and its committees and in the Maritime Planning Agency.

3. Individuals shall be experts and civil servants, appointed or elected in a personal capacity to serve in the Maritime Secretariat or any of its organs or in the Planning Agency.
4. The initial members of the Regime shall be those Member States of the United Nations or any of the specialized agencies which shall have signed this Statute within ninety days after it is opened for signature and shall have deposited an instrument of ratification.
5. Other Members of the Regime shall be those states, whether or not Members of the United Nations or any of the specialized agencies, which deposit an instrument of acceptance of this Statute after application for registration at the Maritime Secretariat. Any dispute over the qualifications of a state shall be referred to the Maritime Court.
6. The Regime is based on the principle of the sovereign equality of all its Members and the full autonomy of all its Associate Members; and all Members, in order to ensure to all of them the rights and benefits resulting from membership, shall fulfill in good faith the obligations assumed by them in accordance with this Statute.

ARTICLE VIII
THE MARITIME ASSEMBLY

A.

The Maritime Assembly shall meet in regular annual session and in such special sessions as shall be convened at the request of the Maritime Commission or a majority of Members and Associate Members. The sessions shall take place at the headquarters of the Regime unless otherwise determined by the Maritime Assembly.

B.

The Maritime Assembly shall consist of five chambers, of eighty-one members each. Members shall serve for three years, except that one third shall be renewed each year.

1. The first chamber shall be elected by the General Assembly of the United Nations with the proviso that
   (a) nine members be elected for each of the nine regions of the world (North America; Latin America; Eastern Europe; Western Europe; the Indian sub-continent; South-East Asia;

Africa South of the Sahara; the Middle East and North Africa; the Far East)

(b) that every Member of the U.N. General Assembly be automatically a candidate for election to the Maritime Assembly, except those not Members of the Regime;

(c) that additional candidates up to a total of twenty-seven for each of the nine regions be nominated by national parliaments or governments or regional parliaments or intergovernmental organizations, including any that may be Members of the Regime but not Members of the United Nations;

(d) that any Member not represented in the first chamber for a three-year period shall have mandatory precedence in the election for the next following Assembly.

For Regions consisting of more than nine Nations the General Assembly thus shall elect alternate Nations for alternate periods, and not all Nations shall be represented in the first chamber at all times. For Regions consisting of less than nine Nations, the General Assembly shall elect more than one, and up to nine, delegate(s) for each Nation, such delegates to be nominated by the national or regional bodies indicated under (c) above.

The eighty-one members of the first chamber shall have each one vote.

2. The second chamber, representing international mining corporations, organizations, unions, producers, and consumers directly interested in the extraction of oil, metals, minerals, and other nonliving resources from the seabed and below the seabed, shall be elected in a manner to be determined by the Secretariat and approved by the first chamber.
3. The third chamber, representing fishing organizations, fish processers and merchants, unions of seamen serving on fishing vessels, consumers, as well as representatives of regional fishing commissions, shall be elected in the above manner.
4. The fourth chamber, representing shipping companies, cable companies and other organizations providing services or communications on or under the oceans, shall be elected in a manner to be determined in the above manner.
5. The fifth chamber, representing scientists in oceanography, marine biology, meteorology, and various other sectors related to the exploration of the seas and the seabed, from intergovernmental and nongovernmental international scientific organizations, regional or universal, shall be elected in the above manner.

C.

1. Each chamber shall elect its own president. The Assembly as a whole shall elect its president and make its own rules of procedure.
2. A majority vote of two chambers—i.e., of the first chamber and the chamber competent in the matter voted upon—shall be required for the adoption of any decision or recommendation. If the two competent chambers fail to agree, they shall discuss the matter in a joint session and vote in common. A simple majority vote of the two joint chambers shall suffice for the adoption of a decision or recommendation.
3. The initiative in proposing decisions, making recommendations and expressing opinions shall be shared equally by all five chambers of the Assembly and by the Commission.
4. Decisions adopted by the Commission shall become effective when approved by two chambers of the Assembly including the first chamber. Decisions adopted by the Assembly shall become effective when passed by the Commission. By a three-fourths majority vote the Commission may return decisions and recommendations to the Assembly where they may not be taken up again before the lapse of a two year period.
5. In any dispute as to which chamber is competent in a matter, the decision of the first chamber of the Assembly shall be final.

D.

The Maritime Assembly may discuss any questions or any matters within the scope of this Statute, issue decisions and recommendations for consideration by the Commission, and give opinions to the membership of the Regime on any such questions or matters.

E.

The Maritime Assembly shall:

1. elect members of the Maritime Commission in accordance with Article IX, A, 2;
2. elect members of the Maritime Planning Agency in accordance with Article X, A, 2;
3. confirm the appointment of the Secretary-General and the Heads of the Secretariats in accordance with Article XI, 1 and 4;
4. confirm the appointment of the Chairman of the Maritime Corpora-

tions and members of their Boards of Directors in accordance with Article XII, 3;

5. make rules for the issuing of licenses for the exploitation of the seabed and for the collection of rents and royalties;
6. make rules for the international operations of multinational corporations and joint ventures;
7. establish criteria for the conservation, development, and exploitation of the living resources of the oceans;
8. make rules and recommendations for navigation, the laying of submarine cables and others uses of ocean space;
9. impose an Ocean Development Tax;
10. review action taken by the Maritime Commission concerning licenses;
11. approve suspension of a Member or Associate Member from the privileges and rights of membership or associate membership;
12. consider the annual report of the Commission;
13. approve the development plan and the budget recommended by the Commission or return it with recommendations as to its entirety or parts to the Commission, for resubmission to the Maritime Assembly;
14. approve the International Mining Code and amendments thereto or return them to the Commission with its recommendations, for resubmission to the Maritime Assembly;
15. review the performance of Members and Associate Members with regard to the International Mining Code;
16. approve reports to be submitted to the United Nations as required by the relationship agreement between the Regime and the United Nations, or return them to the Commission with its recommendations;
17. approve any agreement or agreements between the Regime and the United Nations and other organizations or return such agreements to the Commission with its recommendations, for resubmission to the Maritime Assembly;
18. approve rules and limitations regarding the exercise of borrowing powers by the Commission; approve rules regarding the acceptance of grants to the Regime; and approve the manner in which general funds may be used;
19. approve amendments to this Statute in accordance with Article XVII.

ARTICLE IX
THE MARITIME COMMISSION

A.

The Maritime Commission shall consist of seventeen members and shall be composed as follows:

1. The outgoing Commission (or, in the case of the first Commission, the Committee on the Peaceful Uses of the Sea-Bed and the Ocean Floor Beyond the Limits of National Jurisdiction of the General Assembly of the United Nations) shall designate five Member States for membership on the Commission.
2. The Maritime Assembly shall elect twelve Member States to membership in the Commission, with due regard to equitable representation on the Commission as a whole of developed and developing Nations, maritime and land-locked Nations, and Nations operating under free-enterprise and socialist economic systems.
3. Any Member State not represented on the Commission may appoint an *ad hoc* representative with the right to vote, whenever its own vital interests are directly concerned; but the number of *ad hoc* members at any time shall be limited to four and the final decision regarding their participation rests with the Commission.
4. The members of the Maritime Commission shall serve for three years; they shall be eligible for re-election for the following term of office.

B.

The Maritime Commission shall meet at such times as it may determine. The meetings shall take place at the headquarters of the Regime, unless otherwise determined by the Commission.

C.

The Maritime Commission shall elect its own Chairman and make its own rules for procedure.

D.

In accordance with Article X, the Maritime Commission shall elect one-half of the elective membership of the Maritime Planning Agency.

E.

In accordance with Article XI, it shall appoint a Secretary General.

F.

In accordance with Article XII, the Maritime Commission shall appoint the Chairmen and at least one-half of the members of the Boards of Directors of the Maritime Corporations.

G.

1. Each member of the Maritime Commission shall have one vote.
2. Decisions on the Regime's development plan and budget shall be made by a two-thirds majority of those present and voting. Decisions on other questions, including the determination of additional questions or categories of questions to be decided by a two-thirds majority, shall be made by a majority of those present and voting.
3. Two-thirds of all members of the Commission shall constitute a quorum.

H.

The Maritime Commission shall have authority to carry out the planning, regulatory, and operative functions of the Regime in accordance with this Statute, subject to its responsibilities to the Maritime Assembly as provided in this Statute.

1. The planning function, under the responsibility of the Commission, is entrusted to the Maritime Planning Agency, in accordance with Article X.
2. The regulatory function, under the responsibility of the Commission, is entrusted to the Maritime Secretariats, in accordance with Article XI.
3. The operative function, under the responsibility of the Commission, is entrusted to the Maritime Corporations in accordance with Article XII.

I.

Subject to the rules enacted by the Maritime Assembly, the Maritime Commission is authorized to issue, regulate, supervise, amend, revoke, and enforce licenses to Member States and to Associate Members, for the peaceful and orderly exploration and exploitation of the ocean floor beyond the limits of national jurisdiction and to collect royalties.

J.

At the request of any Member or Associate Member, the Maritime Commission is authorized to issue emergency orders to prevent serious harm to the marine environment arising out of any exploration or exploitation activity and communicate them immediately to parties involved.

K.

The Maritime Commission shall prepare an annual report to the Maritime Assembly concerning the affairs of the Regime. The Commission shall also prepare for submission to the Maritime Assembly such reports as the Regime is or may be required to make to the United Nations or to any other organization the work of which is related to that of the Regime. These reports, along with the annual reports, shall be submitted to Members and Associate Members of the Regime at least one month before the regular annual session of the Maritime Assembly.

L.

The Maritime Commission may establish such committees as it may deem useful. It may appoint persons to represent it in its relations with other organizations.

ARTICLE X
THE MARITIME PLANNING AGENCY

A.

The Maritime Planning Agency shall be composed of economists, scientists, administrators, and other experts, selected as follows:

1. one-half of its elective membership shall be elected by the Maritime Commission;
2. one-half of its elective membership shall be elected by the Maritime Assembly;
3. the members of the Intersecretariat Committee on Scientific Programmes Relating to Oceanography, the members of the Inter-Agency Consultative Board of the U. N. Development Programme,

the President of the World Bank, and the Chairmen of the Maritime Corporations shall be members *ex officio.*

B.

The Maritime Planning Agency shall elect its own chairman, establish its own subcommittees, and adopt its own rules and regulations.

C.

It shall be the responsibility of the Planning Agency to coordinate all efforts and projects presently undertaken by all organizations, within the U.N. system and outside, in the sphere of its competence; to prepare plans to maximize development and exploitation of living and nonliving ocean resources and to ensure their conservation; to prepare a budget for the Regime; to allocate revenue accruing to the Regime from fees, royalties, taxes or grants, and to take appropriate measures to protect developing nations against the fluctuation of prices of minerals and metals and, in general, maximize the creation of wealth from the oceans while minimizing harmful interference with the interests of land-based industries and economies.

D.

Each Member State and each Associate Member and each Regional Committee referred to under Article XIII shall submit each year to the Planning Agency a progress report and development plan to be stored in the Agency's computer and included in the world plan. In integrating the plans, the Planning Agency shall give due consideration to:

1. the usefulness of the plan, including its scientific and technical feasibility;
2. the adequacy of funds and technical personnel to assure its effective execution;
3. the adequacy of proposed health and safety standards;
4. the equitable distribution of financial grants;
5. the special needs of the underdeveloped areas of the world;
6. and such other matters as may be relevant.

E.

The Planning Agency shall make long-range ecological projections and over-all forecasts up to fifty years and beyond; ten-year plans, and annual programs. The long-range projections shall be published every five years. The ten-year plan shall be a general estimate of probable developments; the first ten-year plan shall give form and substance to the International Decade of Ocean Exploration. The annual program shall provide readjustment to developing conditions and fix the annual budget.

F.

The ten-year plan shall be submitted by the Chairman of the Planning Agency to the Maritime Commission, and with the Commission's recommendations, to the Maritime Assembly, one year prior to its going into effect. The annual program shall be submitted to the Commission, to Members and Associate Members, one month prior to the opening of the Regular Annual Session of the Maritime Assembly.

G.

Plans shall be published in every Member State and shall be fully discussed by all chambers of the Assembly and by all interested scientific, economic, and social organizations.

H.

To go into effect, plans must be adopted by the Commission and by the first chamber of the Assembly while the remaining four chambers must adopt only the sections of the plan that concern their respective activities. The first chamber of the Assembly shall determine whether the adoption of a given section by a given chamber is required.

I.

The Chairman of the Planning Agency shall give to the Assembly an annual progress report.

ARTICLE XI
THE MARITIME SECRETARIATS

1. The Maritime Commission, with the approval of the Maritime Assembly, shall appoint a Secretary General who shall be the chief administrative officer of the Regime.
2. The Secretary General shall act in that capacity in all meetings of the Maritime Assembly, the Maritime Commission, and the Mari-

time Planning Agency, and shall perform such other functions as are entrusted to him by these organs.

3. The Secretary General may bring to the attention of the Commission, the Assembly, or the Planning Agency any matter within each organ's competence; he shall bring to the attention of the Commission any matters that in his opinion may constitute a threat to the marine environment.
4. The Secretary General shall establish a Secretariat for Ocean Mining, a Secretariat for Deep-Sea Oil Extraction, a Secretariat for Fisheries and Aquaculture, A Secretariat for Shipping and Communications, a Secretariat for Science and Technology, and others as they may become necessary. The heads of the Secretariats shall be elected by the Maritime Assembly in accordance with Article VIII, E, 3, upon nomination by the Secretary General.
5. The Secretary General shall prepare the slates of candidates for the elections to the Maritime Assembly and the Maritime Planning Agency.
6. In the performance of their duties the Secretary General and the staff shall not seek or receive instructions from any government or from any other authority external to the Regime. They shall refrain from any action that might reflect on their position as international officials responsible only to the Regime.
7. Each Member and each Associate Member of the Regime undertakes to respect the exclusively international character of the responsibilities of the Secretary General and the staff and not to seek to influence them in the discharge of their responsibilities.

ARTICLE XII
THE MARITIME CORPORATIONS

1. The Maritime Commission, with the approval of the Maritime Assembly may establish:
   (a) an *Ocean Science Corporation*, responsible for conducting programs of research development of ocean science and technology; for coordinating national and private programs; for servicing an international ocean data center; and for acting as a repository and clearing house for information;
   (b) an *Ocean Weather Corporation*, responsible for meteorological data gathering, weather forecasting, control and modification and providing services for a fee to nations and corporations;

(c) an *Ocean Petroleum Corporation,* responsible for prospecting, developing, and producing petroleum products from the deep oceans, by itself or in joint ventures with other national or private oil corporations;
(d) an *Ocean Mining Corporation,* responsible for prospecting for minerals, developing underwater recovery methods, and producing minerals from the seabeds, by itself or in joint venture with corporations;
(e) other operative corporations, in accordance with technological and economic requirements.

2. The Corporations are controlled subsidiaries of the Ocean Regime. The Ocean Regime shall advance at least one-half of their capital and elect at least one-half of the members of their boards of directors. The balance of their capitals and boards shall be supplied by those states or public or private corporations who choose to subscribe, subject to the reservation of adequate representation for the developing nations.
3. The Chairmen of the Boards and at least one-half of the Members of the Boards shall be elected by the competent chambers of the Maritime Assembly in accordance with Article VIII, E, 4.
4. The Corporations shall be entitled to representation in the competent functional chambers of the Maritime Assembly.
5. Profits on the Regime's investment in the Corporations' stocks will be returned to the Regime's assets.

ARTICLE XIII
REGIONAL ARRANGEMENTS

A.

Coastal states adjacent to land-locked seas may establish regional organizations to meet the special needs of such areas. There shall not be more than one regional organization in each area.

B.

Each regional organization shall be an integral part of the Regime in accordance with this Statute.

C.

Each regional organization shall consist of a regional committee and a regional office.

D.

Regional committees shall be composed of representatives of Member States and Associate Members, including scientific institutions and organizations.

E.

Regional committees shall meet as often as necessary and shall determine the place of each meeting.

F.

Regional committees shall adopt their own rules of procedure.

G.

The functions of regional committees shall be:

1. to formulate policies governing matters of an exclusively regional character;
2. to supervise the activities of the regional office;
3. to suggest to the regional office the calling of technical conferences and such additional work or investigation and research as in the opinion of the regional committee would promote the objectives of the Regime within the region;
4. to cooperate with the respective regional committees of the United Nations and with those of other specialized agencies and with other regional international organizations having interests in common with the Regime;
5. to advise the Maritime Assembly, the Commission, and the Planning Agency on matters which have wider than regional significance;
6. to recommend additional regional appropriations by the governments of the respective regions if the proportion of the world budget of the Regime allotted to that region is insufficient for the carrying out of the regional functions;
7. to perform such other functions as may be delegated to the regional committee by the Assembly, the Commission, or the Planning Agency.

H.

Subject to the general authority of the Maritime Commission, the regional office shall be the administrative organ of the regional committee. It shall, in addition, carry out within the region the decisions of the Maritime Assembly and the Commission.

I.

The head of the regional office shall be the regional secretary appointed by the regional committee.

J.

The staff of the regional office shall be appointed by the regional secretary in agreement with the regional committee.

ARTICLE XIV
THE MARITIME COURT

1. The function of the Court is to ensure the rule of law in the interpretation and application of the law of the seas, of the present Statute, and of its implementing regulations.
2. The Court shall be composed of eleven Judges, appointed for six years, by agreement among the governments of Member States, from among persons of recognized independence and competence. A partial change in membership of the Court shall occur every three years. No more than two Judges from any region and no more than one Judge from any nation shall sit on the Court at any given time.
3. The number of Judges may be increased by unanimous vote of the Commission on proposal by the Court. The Judges shall designate one of their number as president for a three-year term.
4. The Court shall have jurisdiction over appeals by a Member State or by an Associate Member for the annulment of decisions and recommendations of the Regime on the grounds of lack of legal competence, substantial procedural violations, violation of the Statute or of any rule of law relating to its application, or abuse of power. Associate Members of the Regime shall have the right of appeal on the same grounds against individual decisions and recommendations concerning them, or against general decisions and recommendations that they deem to involve an abuse of power affecting them. The appeals provided for in this Article must be taken within one month from the date of the notification or the publication, as the case may be, of the decision or recommendation.
5. If the Court should annul a decision or recommendation of the

Regime, the matter shall be remanded to the competent organ. The latter must take the necessary measures in order to give effect to the judgment of annulment. In case a decision or recommendation is adjudged by the Court to involve a fault for which the Regime is liable, and causes a direct and particular injury to an enterprise or a group of enterprises, the Commission must take such measures, within the powers granted to it by the present Statute, as will assure an equitable redress for the injury resulting directly from the decision or recommendation that has been annulled, and, to the extent necessary, must grant reasonable indemnity. If the Commission fails to take within a reasonable period the measures required to give effect to a judgment of annulment, an appeal for damages may be brought before the Court.

6. In cases where the Commission is required by a provision of the present Statute or of implementing regulations to issue a decision or recommendation, and fails to fulfil this obligation, such omission may be brought to its attention by the Member States, the Associate Members, or the Maritime Assembly, as the case may be. The same shall be true if the Commission refrains from issuing a decision or recommendation which it is empowered to issue by the provisions of the present Statute or implementing regulations, where such failure to act constitutes an abuse of power. If at the end of a period of two months the Commission has not issued any decision or recommendation, an appeal may be brought before the Court, within a period of one month, against the implicit negative decision which is presumed to result from such failure to act.

7. Prior to imposing a pecuniary sanction or fixing a daily penalty payment, the Commission shall give the interested enterprise an opportunity to present its views. An appeal to the general jurisdiction of the Court may be taken from the pecuniary sanctions and daily penalty payments imposed under the provisions of the present Statute or implementing regulations. In support of such an appeal, and under the terms of paragraph 4 of the present Article, the petitioners may contest the regularity of the decisions and recommendations which they are charged with violating.

8. If a Member State shall deem that in a given case an action of the Regime or a failure by it to act is of such a nature as to provide fundamental and persistent disturbances in the economy of such State, it may bring the matter to the attention of the Commission. Should the Commission recognize the existence of such a situation, it shall decide on the measures to be taken, under the terms of the

present Statute, to correct such situation while at the same time safe-guarding the essential interests of the Regime. When an appeal is taken to the Court under the provisions of the present Article against such decision or against the explicit or implicit decision refusing to recognize the existence of the situation mentioned above, the Court shall review the sufficiency of the grounds of such decision. In case of annulment, the Commission shall decide, within the framework of the Court's judgment, the measures to be taken to fulfill the objectives set forth in this paragraph.

9. Appeals to the Court shall not have the effect of suspending the execution of a decision or recommendation. However, if in the Court's judgment circumstances so demand, the Court may order the suspension of the execution of the decision or recommendation in question. It may prescribe any other necessary provisional measures.
10. Subject to the provisions of paragraph 5, the Court shall have jurisdiction to assess damages against the Regime, at the request of the injured party, in cases where an injury results from a fault involved in an official act of the Regime in the execution of the present Statute law. It shall also have jurisdiction to assess damages against any official or employee of the Regime, in cases where injury results from a personal fault of such official or employee in the performance of his duties. If the injured party is unable to recover such damages from such official or employee, the Court may assess an equitable indemnity against the Regime.
11. Litigation between two or more Member States may be brought before the International Court of Justice by agreement between the Member States.
12. The Court shall have jurisdiction in disputes between the Regime and third parties, whenever both sides have accepted the jurisdiction of the Court for the purpose.
13. When the validity of acts of the Regime is contested in litigation before a national tribunal, such issue shall be certified to the Court, which shall have exclusive jurisdiction to rule thereon.
14. The Court shall have such jurisdiction as may be provided by any clause to such effect in a public or private contract to which the Regime is a party or which is undertaken on its account.
15. The Court shall have jurisdiction in any other case provided for in any supplementary provision of the Statute. It may also exercise jurisdiction in any case relating to the objects of the present Statute, where the laws of a Member State grant such jurisdiction to it.

16. Litigant parties shall have the right of appeal from the determinations of the Maritime Court to the International Court of Justice. Such appeals may be taken by means of a request by the Regime for an advisory opinion, with the litigants stipulating in advance, as a condition of such appeal, to be bound by such advisory opinion.
17. The Court shall have jurisdiction to decide any dispute or controversy as to membership in the Regime or any of its organs.
18. The Code of the Court shall be contained in a Protocol annexed to the present Statute.

ARTICLE XV
DEVICES AND INSTALLATIONS

1. Subject to appropriate regulations prescribed by the Regime, a Member State or Associate Member shall be entitled to construct and maintain or operate on the seabed and below the seabed installations and other devices necessary for the exploration and exploitation of its natural resources and to establish safety zones around such installations and devices and to take in those zones measures necessary for their protection.
2. The safety zones referred to in this Article may extend to a distance of a five hundred meter radius around the installations and other devices which have been erected, measured from each point of their outer edge. Ships of all nationalities must respect these safety zones.
3. Such installations and devices do not possess the status of islands and have no territorial sea of their own.
4. Due notice must be given of the construction of any such installations, and permanent means for giving warning of their presence must be maintained. Any installations which are abandoned or disused must be entirely removed by the Member State or Associate Member responsible for its construction.
5. Neither the installations and devices nor the safety zones around them may be established where interference may be caused to the use of recognized sea lanes essential to international commerce and navigation.
6. All stations, installations, equipment, and sea vehicles, machines, and capsules used on the seabed or below the seabed, whether manned or unmanned, shall be open to representatives of the Regime.

ARTICLE XVI
THE PEACEFUL USE OF THE SEABED AND OF OCEAN SPACE

A.

The seabed and what is below the seabed shall be used for peaceful purposes only.

B.

Each Member State or Associate Member shall accept as binding the 1971 Treaty on the Prohibition of the Emplacement of Nuclear Weapons and Other Weapons of Mass Destruction on the Seabed and the Ocean Floor and the Subsoil Thereof (26025).

C.

The prohibitions of that Treaty shall not be construed to prevent:

1. the use of military personnel or equipment for scientific research or any other peaceful purpose;
2. the use or stationing of any device on the seabed or below the seabed that is designed and intended for purposes of submarine or weapons detection, identification, or tracking.

D.

Any military personnel used for the above purposes must wear the insignia of the United Nations Forces and must report on its activities and findings to the Maritime Commission and to the Security Council of the United Nations.

E.

Any device installed on or under the ocean floor in accordance with paragraph C, 3, above, shall be open to inspection by representatives of the Regime in accordance with Article XV, 6.

F.

Agreements for the further and comprehensive demilitarization of ocean space shall be prepared by the Regime, in consultation between the Disarmament Committee and the Regime, and shall be submitted to Member States for ratification.

G.

1. Existing fishery constabularies and patrols, whether national or co-operating on a regional basis; inspectorates under the responsibility of organizations such as FAO, IMCO, IOC; patrols such as those of the International Cable Protection Committee; inspectorates of multinational insurance companies and other such security bodies, including stand-by units of national coast-guards and navies thus designated by their Nations, shall register with the Secretary General and with the Regional Offices.
2. They shall cooperate and integrate their activities at the regional or global level wherever possible.
3. The Secretary General or the Regional Offices may call on registered forces for a period of training in international peace keeping and for such tasks as the inspection of installations and sea-going equipment, the enforcement of safety standards, the prevention of damages to the ocean environment and, where such damages have occurred, for the assessment of liabilities.

ARTICLE XVII
AMENDMENTS

Any Member State or Associate Member may propose amendments to this Statute. Amendments shall enter into force when approved by a majority of the Assembly and the Commission and ratified by a majority of Member States.

ARTICLE XVIII
HEADQUARTERS OF THE REGIME

The Headquarters of the Regime shall be established on the Island of Malta.

ARTICLE XIX
SIGNATURE, ACCEPTANCE, AND ENTRY INTO FORCE

1. This Statute shall be open for signature on ______, 1973 by all Member States of the United Nations or any of the Specialized Agencies, and shall remain open for signature by those States for a period of ninety days.

2. The signatory States shall become parties to this Statute by deposit of an instrument of ratification.
3. Instruments of ratification by signatory States and instruments of acceptance by States whose membership has been established under Article VII of this Statute shall be deposited with the governments of the United Kingdom, the U.S.A., and the U.S.S.R., hereby designated as Depositary Governments.
4. Ratification or acceptance of this Statute shall be affected in accordance with the respective constitutional processes of the States concerned.
5. This Statute shall come into force when eighteen States have deposited instruments of ratification in accordance with paragraph 2 of this Article.
6. The Depositary Governments shall promptly inform all States signatory to this Statute of the date of each deposit of ratification and the date of entry into force of the Statute. The Depositary Governments shall promptly inform all signatories and members of the dates on which States subsequently become parties thereto.

ARTICLE XX
WITHDRAWALS

Any Member State or Associate Member may give notice of its withdrawal from the Regime one year after its entry into force, by written notification to the Secretary General. Such withdrawal shall take effect one year from the date of receipt of this notification.

ARTICLE XXI
REGISTRATION WITH THE UNITED NATIONS

1. This Statute shall be registered by the Depositary Governments pursuant to Article 102 of the Charter of the United Nations.
2. Agreements between the Regime and any Member or Members, agreements between the Regime and any other organization or organizations, and agreements between Members subject to approval of the Regime shall be registered with the Regime. Such agreements shall be registered by the Regime with the United Nations if registration is required under Article 102 of the Charter of the United Nations.

ARTICLE XXII
AUTHENTIC TEXTS
AND CERTIFIED COPIES

This Statute, done in the Chinese, English, French, Russian, and Spanish languages, each being equally authentic, shall be deposited in the archives of the Depositary Governments. Duly certified copies of this Statute shall be transmitted by the Depositary Governments to the governments of the other signatory States, to the governments of States admitted to membership under Article VII, and to the executive organs of Associate Members.

In witness whereof the undersigned, duly authorized, have signed this Statute.

Done at the Headquarters of the United Nations, this tenth day of March, nineteen hundred and seventy-three.

Notes

CHAPTER 3

1. *Childe Harold's Pilgrimage,* Canto 4, Stanza 179.
2. Burton, I., R. Kates, and R. Snead, *The Human Ecology of Coastal Flood Hazard in Megalopolis,* Research Paper No. 115 (Chicago: Department of Geography, University of Chicago, 1969).

CHAPTER 4

1. Ecological Society of America, *Ad Hoc* Weather Working Group, Biological aspects of weather modification, Ecological Society of America, Bulletin, Vol. 47 (March 1966), pp. 39-78.
2. Sewell, W. R. Derrick (ed.), *Human Dimensions of Weather Modification,* Department of Geography Research Paper No. 105. (Chicago: University of Chicago Press, 1966).
3. Battan, Louis J., "Weather Modification in the U.S.S.R.–1969" *Bulletin of the American Meteorological Society,* Vol. 50, No. 12 (Dec., 1969.) pp. 924-945.
4. Landsberg, H.E., "Climatic Consequences of Urbanization," Paper for AAAS Symposium, *Climate and Man* (Boston, Dec., 1969). Mimeo.
5. Taubenfeld, Howard J., "The International Lawyer and Weather Modification," Appendix IX in *Human Dimensions of the Atmosphere,* National Science Foundation (Washington, —), p. 99.
6. Ostrom, Vincent. 1968. "Needs for Research on the Political Aspects of the Human Use of the Atmosphere." Appendix VI in *Human Dimensions of the Atmosphere,* National Science Foundation, (Washington, —), p. 71.
7. Taubenfeld, Rita F. and Howard J., "Some International Implications of Weather Modification Activities," *International Organization,* Vol. 23, No. 4 (1969), pp. 808-33.
8. Ibid.

CHAPTER 7

1. K. O. Emery, *An Oceanographer's View of the Law of the Sea,* Ist. Affari Intern., Symp. Intern. Regime Seabed (Rome, 30 June–5 July 1969). In press.
   A. J. Guilcher, *The Configuration of the Ocean Floor and its Subsoil:* Geopolitical Implications, 1st. Affari Intern. Symp. Intern. Regime Seabed (Rome, 30 June–5 July 1969). In press.
   J. L. Worzel, Ch. 8 in *Geology of Shelf Areas,* D. T. Donovan, Ed., (Edinburgh: Oliver & Boyd, 1968).
   V. F. McKelvey and F.F.H. Wang, *Discussion to Accompany Geologic Investigations Map 1-632,* U. S. Geol. Survey (1969).
   V. F. McKelvey, J.F. Tracey, G.E. Stoertz, and J.E. Vedder, *U.S. Geol. Survey Circular,* No. 619 (1969).
   H. W. Menard and S.M. Smith, *Journ. Geophysical Res.,* Vol. 17, No. 8 (1966), pp. 4305-4325.
2. Emery, op. cit.
3. Worzel, op. cit.
4. Menard and Smith, op. cit.
5. McKelvey and Wang, op. cit.
6. McKelvey et al., op. cit.
7. *Mineral Resources of the Sea,* Rept. of the Secretary-General, U.N. Doc. E/4680 (1969).
   Preston Cloud, Ch. 7, *Mineral Resources from the Sea,* supra, note 2, pp. 135-155.
8. Harold L. James, Proc. *Symp. on Mineral Resources of the World Ocean,* U. of Rhode Island, Grad. School of Oceanog. Occ. Paper No. 4 (1968), pp. 39-44.
9. *Resources of the Sea,* Report of the Secretary-General, U.N. Docs. E/4449, E/4449/Add. 1, E/4449/Add. 2 (1968).
   Mineral Resources of the Sea, op. cit.
   Cloud, op. cit.
10. Cloud, op. cit.
11. James, op. cit.
12. Cloud, op. cit.
13. Cloud, op. cit.
14. Cloud, op. cit.
    James, op. cit.
15. W. E. Ricker, Ch. 5, *Food from the Sea,* supra note 2, pp. 86-108.
16. *Mineral Resources of the Sea,* op. cit.
    Resources of the Sea, op. cit.
17. McKelvey and Wang, op. cit.
    Cloud, op. cit.
    Resources of the Sea, op. cit.
18. McKelvey and Wang, op. cit.

19. Center for the Study of Democratic Institutions, *Pacem In Maribus,* prospectus (1969).
20. L.G. Weeks, *Jour. Petr. Technology* (April 1969), pp. 377-385.
21. Ibid.
22. M.K. Hubbert, Ch. 8, *Energy Resources,* supra note 2, pp. 157-242.
23. Weeks, op. cit.
24. Ibid.
25. *Resources of the Sea,* op. cit.
26. *Mineral Resources of the Sea,* op. cit.
    Resources of the Sea, op. cit.
27. McKelvey and Wang, op. cit.
    McKelvey, Tracey et al., op. cit.
    Menard and Smith, op. cit.
    Resources of the Sea, op. cit.
    Committee on Petroleum Resources Under the Ocean Floor, *Petroleum Resources Under the Ocean Floor* (Washington: National Petroleum Council, 1969).
28. McKelvey and Wang, op. cit.
29. Committee on Petroleum Resources Under the Ocean Floor, op. cit.
30. Ibid.
31. *Mineral Resources of the Sea,* op. cit.
32. Committee on Petroleum Resources Under the Ocean Floor, op. cit.
33. Ibid.
    McKelvey and Wang, op. cit.
34. Ibid.
35. Committee on Petroleum Resources Under the Ocean Floor, op. cit.
36. Ibid.
37. Ibid.
38. *Prospects for Oil Shale Development, Colorado, Utah, and Wyoming* (U.S. Dept. of Interior, May 1968), p. B-19.
39. Committee on Petroleum Resources Under the Ocean Floor, op. cit.
    Committee on Deep Sea Resources of the American Branch of the International Law Association, *Interim Report* (1968). See also discussion by National Petroleum Council, supra, note 36, pp. 55-67.
40. T. F. Gaskell, *Oil and Natural Gas: Exploration, Evaluation and Exploitation of Deep Water Petroleum,* Ist. Affari Inter., Symp. Intern. Regime Seabed (Rome, 30 June–5 July, 1969). In press.
41. *Catches and Landings* 1968. F.A.O. Yearbook of Fishery Statistics, Vol. 26 (Rome, 1969).
42. Committee on Petroleum Resources Under the Ocean Floor, op. cit.
    F.A.O. Conference on Investment in Fisheries, Fishing News International, Vol. 8, No. 11 (1969), pp. 18-24.
    W. E. Ricker, Ch. 5, *Food from the Sea,* supra note 2, pp. 86-108.
43. Center for the Study of Democratic Institutions, op. cit.

44. M.B. Schaefer and R. Revelle, Ch. 4 in *Natural Resources,* M.R. Huberty and W.L. Flock, Eds. (New York: McGraw-Hill, 1959).
45. Convention on Fishing and Conservation of the Living Resources of the High Seas, 17 U.S.T. 138, T.I.A.S. 5969 (April 29, 1958, in force March 20, 1966).
46. See, for example, F.T. Christy and A. Scott, *The Common Wealth in Ocean Fisheries,* Resources for the Future (Washington, 1965).
47. M.B. Schaefer and D.L. Alverson, *World Fish Potentials,* U. of Washington Publ. in Fish., N.S. Vol. 4 (1968), pp. 81-85.
48. Ricker, op. cit.
49. *F.A.O. Conference on Investment in Fisheries,* op. cit.
50. M.B. Schaefer, *Trans. American Fish.* Soc., Vol. 94, No. 2 (1965), pp. 123-128.
51. Ricker, op. cit.
52. O.I. Koblenz-Mishke, *Okeanologiia,* Vol. 5, No. 2 (1965), pp. 325-337.
53. O.I. Koblenz-Mishke, V.V. Volkovinsky, and J.G. Kabanova, *Plankton Primary Production of the World Ocean,* I.C.S.U., Scientific Committee on Ocean Research, Symposium on Scientific Exploration of the South Pacific (1969). In press.
54. Schaefer, op. cit.
55. *F.A.O. Conference on Investment in Fisheries,* op. cit.
56. Schaefer and Revelle, op. cit.
57. Convention on Fishing and Conservation of the Living Resources of the High Seas, op. cit.
58. Ricker, op. cit.
59. *F.A.O. Conference on Investment in Fisheries,* op. cit.
60. Convention on Fishing and Conservation of the Living Resources of the High Seas, op. cit.

CHAPTER 8

1. Although no reliable estimates are available, the total value of all other mineral products taken from the sea is rather insignificant compared with those of fish and oil, and the situation is unlikely to change in the foreseeable future.
2. Assuming, in other words, that we shall be harvesting forms of animals more or less similar to those already under exploitation and that methods of harvesting will not be radically different from those employed now.
3. For biological and technological reasons, the utilization of the main forms of marine plants, phytoplankton, as a major source of food is out of the question.
4. At present over one-third of the fish taken from the sea is processed into fishmeal which is used for feeding domestic animals, particularly poultry, and the proportion is likely to increase further.

5. As of 1968, for example, claims included: Argentina (two hundred miles), Cambodia (shelf to fifty meters including superjacent waters), Chile (two hundred), Ecuador (two hundred), El Salvador (two hundred), Guinea (130), Indian (one hundred for conservation purposes), Republic of Korea (varying distances), Nicaragua (two hundred), Panama (two hundred), Peru (two hundred), Philippines and Indonesia claim waters within their respective archipelagos. (From Marine Science Affairs—Report of the President to the Congress on Marine Resources and Engineering Development, Washington, D. C., 1969).
6. The Arabian Sea, for example, is often referred to as one of the potentially richest fishing areas of the world ocean, but our knowledge of its resources is extremely poor.

CHAPTER 9

1. *Commission on Marine Science, Engineering, and Resources* (Stratton), (Washington, D.C.: U.S. Government Printing Office, 1969), p. 151.
   V.E. McKelvey, J.I. Tracy Jr., George E. Stoertz, and John G. Vedder, "Subsea Mineral Resources and Problems Related to Their Development," U.S. Geological Survey Circular 619 (1969).
2. Carl F. Austin, "In the Rock, a Logical Approach for Undersea Mining of Resources" in *Engineering and Mining Journal,* Vol. 168, No. 8 (August 1967).
3. Preston Cloud, "Mineral Resources from the Sea" in *Resources and Man,* National Academy of Sciences, National Research Council (1968), p. 135.
4. Harold L. James, *Proceedings of the Symposium on Mineral Resources of the World Ocean,* University of Rhode Island, Graduate School of Oceanography, OCE Paper No. 4 (1968), pp. 39-44.
5. E. D. Goldberg, "The Oceans as a Chemical System" in *The Sea,* N. Hill, Ed., Vol. 2 (Wiley, 1963), pp. 3-25.
6. A.R. Miller, D.C. Densman, E.T. Degens, J.C. Hathaway, F.T. Mannheim, P.T. McFarlin, R. Pocklington, A. Jodela, "Hot Brines and Recent Iron Deposits in Deeps of Red Sea" in *Geochimica et. Comochimica Acta,* Vol. 30 (1966), pp. 341-349.
   V.E. McKelvey and F.H. Wang, "Discussion to Accompany Geological Investigations Map 1-632," U.S. Geological Survey (1969).
7. McKelvey, Tracy, et al., op. cit.
   James, op. cit.
8. McKelvey, Tracy et al., op. cit.
9. John L. Mero, *The Mineral Resources of the Sea* (New York: Elsevier Publishing Co., 1965).
10. McKelvey and Wang, op. cit.
11. Ibid.
12. McKelvey, Tracy et al., op. cit.

Philip E. Sorensen and Walter J. Mead, "A Cost-Benefit Analysis of Ocean-Mineral Resource Development: The Case of Manganese Nodules" in *American Journal of Agricultural Economics,* Vol. 50, No. 5 (December 1968), pp. 1611-1620.
13. Mero, op. cit.
14. McKelvey and Wang, op. cit.
15. Sorensen and Mead, op. cit.
16. V.E. McKelvey, "Progress in the Exploration and Exploitation of Hard Minerals from the Sea Bed," U.S. Geological Survey Circular 619 (1969).
17. Ibid.

CHAPTER 10

1. Public Law 89-454.
2. United Nations General Assembly Resolution 2172 (XXI) of December 6, 1966.
3. See United Nations General Assembly Resolutions previously referred to, as well as 2413 (XXIII) of December 17, 1968, on Living Resources; 2574A-D (XXIV) of December 15, 1969; 2566 (XXIV) of December 13, 1969, on Marine Pollution; and 2606F (XXIV) of December 16, 1969, on Seabed Disarmament.
4. United Nations General Assembly Resolution 2574C (XXIV) of December 15, 1969.

CHAPTER 12

1. See William T. Burke, "Statement on Some Policy Issues Relating to the Continental Shelf and Other Matters—For: Special Subcommittee on the Continental Shelf, Senate Committee on Interior and Insular Affairs—" (January 22, 1970), mimeo.
2. Stuart Scheingold, transcript of remarks at *Pacem in Maribus* Preparatory Conference on the Legal Framework and Continental Shelf (University of Rhode Island, January 30–February 1, 1970) (Center for the Study of Democratic Institutions, Santa Barbara).
3. See Christy, "Fisheries Goals and the Rights of Property," *Transactions of the American Fisheries Society,* Vol. 98, No. 2 (April, 1969), p. 373 ff. See also Christy, "Marigenous Minerals: Wealth, Regimes, and Factors of Decision" in *Proceedings of the Symposium on the International Regime of the Sea-Bed* (Rome: Istituto Affari Internazionali), in press.
4. M.B. Schaefer, "Fisheries Productivity," *School Science and Mathematics* (February, 1969), p. 145.
5. See J.E. Carroz, "Living-Resource Management: Regional Fishery Bodies," paper prepared for this Conference.

6. Oda, "Japan and International Conventions Relating to North Pacific Fisheries," 43 *Washington Law Review* 67 (Washington, 1967), p. 72.
7. International Panel, Commission on Marine Science, Engineering and Resources, "An International Legal-Political Framework for Exploiting the Living Resources of the High Seas" in *Our Nation and the Sea,* Vol. 3 (Washington: U.S., GPO, January, 1969), p. VIII-63.
8. Oda, op. cit., p. 65.
9. "Korea Welds Opinions of Commission," *National Fisherman* (January, 1970), p. 15-A.
10. Ambassador McKernan of the U.S. refers to these as "Executive Agreements, because for the most part they are less formal than the regular fisheries conventions. They have not had Senate ratification and so, in general at least, they have not had the broad public debate that occurs with the more formal agreement or convention." McKernan, "International Fisheries Arrangements Beyond the Twelve Mile Limit," in Alexander, ed., *The Law of the Sea: International Rules and Organization for the Sea* (Kingston: University of Rhode Island for the Law of the Sea Institute, March 1969), p. 255.
11. James Crutchfield and Giulio Pontecorvo, *The Pacific Salmon Fisheries: A study of Irrational Conservation* (Baltimore: Johns Hopkins Press for Resources for the Future, Inc., 1969).
12. Edward Lynch, Richard Doherty, and George Draheim, *The Groundfish Industries of New England,* U.S. Fish and Wildlife Service (Washington, D.C. 1961), circ. p. 121.
13. *Report of the Committee of Inquiry into the Fishing Industry* (London: H.M.S.O., January, 1961), p. 27.
14. Virgil Norton, "Some Potential Benefits to Commercial Fishing Through Increased Search Efficiency: A Case Study—The Tuna Industry" (University of Rhode Island, 1969).
15. A model for such an approach can be found in McDougal and Burke, *The Public Order of the Oceans* (New Haven: Yale University Press, 1962).
16. Ibid., p. ix.

CHAPTER 14

1. In this paper, I have intentionally avoided discussions of the labels used to characterize the European Community and the integrative process. The term neo-functional would reflect my own position more accurately than does functional. But whether the Community is functional, neo-functional, supranational etc., just what each of these terms means, and whether there have been or are now other such organizations need not be considered here. While interesting problems, they do not bear directly on the relevance of the European Community for an Ocean Regime. For an exhaustive analysis of the concept of supranationalism,

see Peter Hay, *Federalism and Supranational Organizations* (Champagne-Urbana: University of Illinois Press, 1966). My own understanding of the distinction between functional and neo-functional approaches is best expressed in Leon N. Lindberg and Stuart A. Scheingold, *Europe's Would-Be Polity: Patterns of Change in the European Community* (Englewood-Cliffs, N.J.: Prentice Hall, 1970), pp. 4-14.

2. As I have noted elsewhere, "Where other parties were in power and/or the effects of the war did not seem so crushing, the integration option was not so appealing—in particular, in Great Britain and Scandinavia." Stuart A. Scheingold, "Domestic and International Consequences of Regional Integration," *International Organization* (forthcoming), note 8.
3. There were also political, social, and diplomatic facets to this expansive future, Ibid.
4. See, for example, Stanley Hoffman, "Obstinate or Obsolete: The Fate of the Nation-State and the Case of Western Europe" reprinted in Joseph S. Nye, Jr., *International Regionalism* (Boston: Little, Brown, 1968), pp. 185-198, and Ellen Frey-Wouters, "The Progress of European Integration," *World Politics*, Vol. 17, No. 3 (April, 1965).
5. For some interesting observations on the intimate interrelationships between crisis and regional integration, see Philippe C. Schmitter, "A Revised Theory of International Integration," *International Organization* (forthcoming).
6. Economists distinguish among customs unions, common markets, and economic unions as increasingly intensive forms of economic mergers, but these categories were not relevant, as such, to Monnet's preconditions. That is, to say, the exact form of the final project was less important than the willingness to move beyond free trade.
7. As quoted in Stefan A. Riesenfeld and Richard M. Buxbaum, "N.V. Algemene Transport—en Expeditie Onderneming Van Gend & Loos c. Administration Fiscale Neerlandaise: A Pioneering Decision of the Court of Justice of the European Communities," *American Journal of International Law*, Vol. 58, No. 1 (January, 1964), p. 155.
8. Costa v. E.N.E.L. as quoted in Andrew Wilson Green, *Political Integration By Jurisprudence* (Leyden: A. W. Sijthoff, 1969), p. 180.
9. See Scheingold, *The Rule of Law in European Integration*, Chapter 14, in particular.
10. André M. Donner, *The Role of the Lawyer in the European Communities* (Evanston, Illinois: Northwestern University Press, 1968), p. 62.
11. For a more systematic explanation of this approach to system growth and a more detailed analysis of the growth of the European Community, see Lindberg and Scheingold, *Europe's Would-Be Polity*, op. cit., pp. 65-82.
12. This was the argument developed by Haas in *The Uniting of Europe* (Stanford, California: Stanford University Press, 1958).

13. Lindberg and Scheingold, *Europe's Would-Be Polity,* pp. 75-76.
14. Stanley Hoffman, "Europe's Identity Crisis," *Daedalus,* 93 (Fall, 1964), p. 1275.
15. The leadership question is considered at some length in Lindberg and Scheingold, *Europe's Would-Be Polity,* particularly pp. 128-33.
16. See Francis T. Christy, Jr., "Fisheries: Common Property, Open Access, and the Common Heritage," *Pacem in Maribus* Preparatory Conference on the Legal Framework and the Continental Shelf.

CHAPTER 19

1. Samuel Feldman: The U.S. Navy D.S.Vs; paper given at Oceanology '69 Conference, 1969. The first of these, the Deep Submergence Rescue Vehicle, has started sea trials. It has a crew of three and can transport twenty-four men at a time. The vessel is "unclassified" and other navies will be able to install an appropriate fixing tube on their submarines. *Ocean Information* (January–February, 1970), p. 4.
2. *Aviation Week & Space Technology* (March 9, 1970), p. 212.
3. See e.g. *Air Force Space Digest* (February, 1970), p. 34.
4. See e.g. Moscow Radio, February 25, 1970 for details of "the new generation of trawlers which are virtually floating factories." The Polish Press Agency (March 2, 1970) reports a new Soviet method for catching fish without nets: they are "enticed by light and extracted with the aid of a special pump installed on the ship." Speeds of one hundred knots are forecast for air cushion vehicles; tankers with a dead-weight of a million tons are being considered; and one of 420,000 tons was built in Japan in 1971.
5. See e.g. *Report on Marine Science & Technology,* H.M. Stationery Office, 1970, Cmnd. 3992, p. 17, paragraph 68.
6. One such storage unit is in operation 58 miles offshore from Dubai, in the Persian Gulf. It is in 160 feet of water, 205 feet high, and 270 feet across. Its capacity is 500,000 barrels and it is equipped with foghorns and warning lights. *Ocean Industry* (September, 1969), p. 9.
7. General Dynamics is said to be proposing a 170,000 ton submarine tanker, 270 meters long, 42 meters across and 25.5 meters draught, with a speed of 18 knots. *Oceans Information* (January–February, 1970).
8. It was reported that Messrs. Deepsea Ventures of Houston, Texas, proposed to demonstrate a collection system, by way of suction pipes, in summer, 1970 *Oceans Information* (December, 1969), p. 6, and (January–February, 1970), p. 8. Television tubes that amplify light by a factor of 30,000 allow the bottom to be inspected in detail—Willard Bascom, "Technology & the Ocean" in *Scientific American* (September, 1969), p. 200.
9. Surveying started in 1970 in Japan for offshore sites suitable for build

ing floating or submarine nuclear power stations, *Kyodo* (January 27, 1970). An artificial island is being constructed in the Black Sea as a base for oil prospecting; it is to be capable of withstanding ice and waves fifteen meters high. Moscow Home Service (November 22, 1969).

10. *Aviation Week & Space Technology* (March 9, 1970), p. 209.
11. *Aviation Week & Space Technology* (January 5, 1970).
12. *Technology Week* (June 12, 1967).
13. Settled March, 1970.
14. *International Herald Tribune* (March 4, 1970).
15. *Financial Times* (March 24, 1970).
16. *Financial Times* (March 5, 1970).
17. The *London Times* (February 28, 1970). On March 25, 1970 it was announced from Brazilia that Brazil's territorial waters had been extended to two hundred miles from the previous twelve.
18. The Soviet Union, several Eastern European countries, Indonesia, Turkey, and Pakistan require such advance permission. See: A. Kobodkin, *Territorial Waters and International Law; International Affairs,* No. 8 (Moscow, 1969), pp. 78 ff.
19. The Antarctic Treaty, that area being of no very great military significance, was not the product of alarm.
20. *Third Report,* Disarmament Commission 1956. PV 58 *et seq*. A rival in foresight, but into the arms race itself, was Hanson Baldwin of *The New York Times,* who on August 13, 1945 (eight days after the Atomic bomb was dropped on Hiroshima) forecast the "triumph of pushbutton war," and the day when Russians as well as Americans would have "long range rockets with atomic warheads . . . that can span oceans and demolish cities."
21. Information derived from The Military Balance 1969-70. The Institute for Strategic Studies; *Aviation Week and Space Technology* (March 9, 1970).
22. USIS Official Text (February 24, 1970), p. 6.
23. *Aviation Week and Space Technology* (March 9, 1970).
24. In a recent oil spill in the Gulf of Mexico, the company concerned, a subsidiary of Standard Oil of California, is said to have admitted not having safety valves as required by United States Federal regulations. *International Herald Tribune* (March 14-15, 1970). See R. & F. Murray, below.
25. The kind of thing that springs to mind is a Santa Barbara-type blowout occurring on the Shelf of one of the smaller West African or Southeast Asian states.
26. E.g., through the "Tanker Owners Voluntary Agreement Concerning Liability for Oil Pollution" (TOVALOP), for the oil carrying industry, which is likely to develop certain disciplinary functions.
27. An unconfirmed report in the *Financial Times* (December 13, 1969) suggests that the British Insurance market will not be willing to accept

third-party risks following oil pollution by tankers beyond a certain definite limit. Even to this point, insurers are likely to insist on the meticulous observance of stringent conditions. For checking that they are being observed, a regular inspectorate may develop.

28. "Agreement for Cooperating in Dealing with Pollution of the North Sea by Oil" (1969).
29. The *London Times* reported, Feb. 4, 1970, professional disagreements over the two-way traffic scheme introduced two years ago in the English Channel. IMCO is said to be preparing an alternative scheme.
30. See Footnote 26.
31. As already happens in the Gulf of Mexico.
32. It is necessary to consider here not only the area beyond that referred to in Article I of the 1958 Convention on the Continental Shelf, but also that of the Continental Shelf itself, if only because the sovereignty of the coastal state is there less than complete, see IV, p. 21. The draft Seabed Treaty of October 30, 1969, refers to the whole area beyond the twelve mile limit of the contiguous zone, but the U.S. draft of May 1969 referred to a three mile limit.
33. ENDC/PV 410 (May 13, 1969), p. 4.
34. Dr. John Craven, Santa Barbara, January 7-9, 1970.
35. There may be some increase in this capability: in project GLORIA, a long-range Side Scan Sonar has obtained "recognizable returns from the deep ocean floor at a distance of ten miles." *Nature* (September 20, 1969), p. 1256.
36. See Chapter 21.
37. *War/Peace Report* (October 1969), p. 7.
38. CCD/269 (October 9, 1969).
39. The English-language *Soviet News*, published by the Press Dept. of the Soviet Embassy in London uses the word "stationing" (November 18, 1969).
40. CCD PV/444 (October 21, 1969), p. 44.
41. Such as the "Deep Submergence Rescue Vehicle" and the "Deep Submergence Search Vehicle" now under development in the U.S. There is no information about equivalent Soviet vehicles but a requirement for such vehicles is imperative, if only for the all-important reason of crew morale in nuclear submarines.
42. Santa Barbara, January 1970.

CHAPTER 20

1. J.S. Cowie, *Mines, Mine-layers, and Mine-laying* (Oxford: Oxford University Press, 1949).
2. Jane's *Fighting Ships: 1945*, Sampson Low, (Marston, 1945).
3. See the author's "Deterrence From the Sea," *Survival* Volume XII, No. 6, June 1970.

4. E.g. *Daily Telegraph*, Volume 23, Number 3, p. 61.
5. Although they contravene the 1907 Hague Convention, flotation mines have occasionally been resorted to. See, for example, M.W. Cagle and F.A. Mason "Wonsan: Battle of the Mines," United States Naval Institute Proceedings (June, 1957).
6. E.g. the 1968-9 edition.
7. Mark Schneider "SABMIS and the Future of Strategic Warfare," United States Naval Institute Proceedings (July, 1969).

CHAPTER 21

1. See also: F.A.O. Limits and Status of the Territorial Sea, Exclusive Fishing Zones, Fishery Conservation Zones and the Continental Shelf. F.A.O. Legislative Series No. 8 (Rome, 1969).
2. The use of fire-power has been known. In 1961 an Aberdeen Trawler off the Faroes kidnaped a Danish fishery patrol boarding party at which the Danish patrol vessel opened fire. See: U.K. Treaty Series Command 1575, No. 117 (1961).
3. For a detailed discussion of this dispute see: Morris Davis, *Iceland Extends its Fisheries Limits* (Universtitsforlaget, 1964).
4. *Report of the Bureau of Commercial Fisheries for the Calendar Year 1967* (Washington, 1969), p. 19.
5. J.E. Carroz and A.G. Roche, "The International Policing of High Seas Fisheries" in *The Canadian Yearbook of International Law 1968*, p. 85, note 102.
6. As of February 1970, oysters were being cultivated in thirteen places in England and Wales, mussels in some ten places, and clams in two places. For the regulations governing this cultivation see Sea Fisheries (Shellfish) Act 1967. For discussion of the potential of mariculture see: J.H. Ryther and G.C. Mathiessen, "Aquaculture, its Status and Potential" in *Oceanus* Vol. XIV, No. 4 (February, 1969).
7. *The Times* (March 3,1970).
8. National laws differ, but the substance of many of them can be gathered from the provisions of the International Convention for the Safety of Life at Sea, 1960, Command 2812, Treaty Series No. 65 (1965) which by July, 1965 has been accepted by thirty countries.
9. Not all lighthouse authorities are publicly operated. The Middle East Navigation Aid Service (formerly the Persian Gulf Lighting Service) provides lighthouses, decca, and navigational aids, and is funded by levy per barrels of oil on oil companies in the Gulf.
10. In Britain there is no unified authority. Apart from the lighthouse and pilotage authorities, the Admiralty issues a weekly notice to mariners containing information about hazards, changes in markings and so on, supplemented by radio broadcasts for urgent information. The Board of Trade issues a numbered series of "M" notices which are of an advisory

nature, and cover such matters as the siting of compasses, newly recommended precautions against fire, shifting cargoes etc. See: D.B. Foy, *Officer Manning—The Neglected Variable in Marine Insurance,* Mimeo. (September, 1968), p. 8.

11. International Convention for the Safety of Life at Sea. op. cit. Chapter 5, Regulation 8, "North Atlantic Routes." pp. 316-8. The regulation referred to is in the process of amendment within IMCO but the final version is not yet available. See also: Chamber of Shipping of the United Kingdom, Annual Report (1968), p. 88.
12. The many conventions prior to 1963 concerning the employment, welfare and status of seamen, are conveniently gathered together in N. Singh, *International Conventions of Merchant Shipping,* Part II (Stevens, 1963), pp. 877-1044.
13. Ibid., pp. 1032-1036.
14. Ibid., pp. 1009-1013.
15. Details of facilities for disposal of waste at oil terminals in a variety of countries are given in: *I.M.C.O. Facilities in Ports for the Reception of Oil Residues. Results of an Inquiry made in 1963* (London, 1964).
16. For Canada and Romania, see: I.M.C.O. Pollution of the Sea by Oil. pp. 93 and 100. U.K. information from Board of Trade private communication.
17. However, many of these prosecutions were for offenses in harbors where both the ship and the evidence were readily available.
18. Singh, op. cit., p. 1150.
19. By early 1968 fifteen states had issued permits for exploration activity beyond the two hundred meter isobar mentioned in the Geneva Convention. The U.S.A. has granted a phosphate lease forty miles off the California coast in the Forty Mile Bank area in 240-4000 feet of water, and oil and gas leases thirty miles off the Oregon coast in fifteen hundred feet of water. Australia has granted permits for up to two hundred miles, and Nicaragua and Honduras for up to 225 miles. See: *Towards a Better Use of the Oceans. A Study and Prognosis,* SIPRI (Stockholm, 1968), p. 26.
20. The Geneva Convention on the Continental Shelf recognizes the rights of States to establish safety zones around exploration and exploitation devices up to a distance of five hundred meters, and "to take in those zones measures necessary for their protection." See: *Report on the First United Nations Conference on the Law of the Sea,* H.M.S.O., (1958). Annex IV, Article 5, paras 2 and 3.
21. A good example of national code is that governing oil and gas operations in the Norwegian area of the continental shelf. See: Government of Norway. Regulations relating to Safe Practice etc. in Exploration for and Exploitation of Petroleum Resources of the Sea-Bed and its Subsoil. Royal Decree of August 25, 1967 (English translation. U.N.A/AC 135/1/Add. 1, March 12, 1968).

22. See: I.C.E.S. Report of the I.C.E.S. Working Group on Pollution of the North Sea. op. cit., p. 4. The principal responsibility for enforcing the legal measures of the Dutch Mining Act for the Continental Shelf is vested in State Inspection Service for Mines. Inspectors have the right to see the documentation of the licensee, to enter all establishments, ships, and aircraft used in operations, and to suspend operations until the code is obeyed.
23. Resolution XVI, 90, "Declaratory Statement of Petroleum Policy in Member Countries," adopted in the XVI Conference of the Organization of Petroleum Exporting Countries in June, 1968, and, "A Pro-forma Regulation for the Conservation of Petroleum Resources" put before the XVII Conference of O.P.E.C. in November, 1968 which decided that it should be adopted in Member Countries.
24. *Report on the First United Nations Conference on the Law of the Sea*, Annex II, articles 2, 26, 27, 28, 29, and 30, H.M.S.O. (1958).
25. The Convention for the Protection of Submarine Cables is reprinted in Singh, op. cit., pp. 275-278. See particularly Article 10.
26. A copy of the proposed articles for the preliminary draft convention on the Legal Status of Ocean Data Acquisition Systems is printed in UNESCO (IOC), *Summary Report of the Third Meeting of the IOC Group of Experts on the Legal Status of Ocean Data Acquisition Systems.* SC/IOC/EG-1/7, SCE/9/89M-ODAS (Paris, December 20, 1969). IOC have also published a useful summary of national and international legislation relevant to O.D.A.S. in IOC, "Legal Problems Associated with Ocean Data Acquistion Systems: A Study of Existing National and International Legislation," SC.69/XVI.
27. Summary Report of the Third Meeting of the IOC Group of Experts. op. cit., Annex III. "Problems to be Resolved in Clarifying the Legal Status of O.D.A.S., and their Solutions." p. 6.
28. Ibid., p. 7.
29. The Radio Regulations, Geneva 1959, are reprinted in Singh, op. cit. pp. 350-629. The French authorities have not only taken action with respect to wireless licenses, but have suspended radiotelegraph and radio telephone services with all ships operating a broadcasting service from outside national territorial limits. This measure has been applied to all types of correspondence, whether incoming or outgoing with the exception of messages relating to the safety of life or of navigation.
30. Recommendation No. 16, Relating to the Measures to be Taken to Prevent the Operation of Broadcasting on Board Ships or Aircraft outside National Territories in Singh, op. cit., pp. 600-601.
31. The U.S.A. tends to fine carriers one thousand dollars for bringing into the country a passenger without a U.S. visa, down to one hundred dollars for every mistake in the passenger manifest. Western European countries do not operate such a system, though many of them require the carriers to pay for the cost of board while a passenger is being in-

vestigated and to ship away passengers refused entry. In spite of these penalties many carriers in and to Europe do not inspect a prospective traveller's papers, though they will tend to when the journey is to the U.S.A.

32. Most customs partols are either armed (the Italian patrol for example) or have access to arms (the British customs patrols are unarmed but have the right to call on the navy, though this has never been done since the 1952 Customs and Excise Act came into force).
33. The seaborne inter-inspection-post patrols are themselves coordinated with land-based patrols. Thus the U.K. coast is separated into areas manned by Customs Coast Preventive Officers who link closely with the police and with the Coast Guards: they operate mainly with cars and shortwave radio. In Hong Kong police patrol the coastline.

Selected Bibliography (1969)

OFFICIAL PUBLICATIONS AND DOCUMENTS

"Ad Hoc Committee on Peaceful Uses of the Sea-Bed Begins Second Session." *U. N. Monthly Chronicle* 5 (1969): 46-49.

Global Ocean Research: A Report. Prepared by the Joint Working Party on the Scientific Aspects of International Ocean Research, nominated by the Advisory Committee Marine Resources Research of the Food and Agriculture Organization, the Scientific Committee on Oceanic Research of the International Council of Scientific Unions, and the Advisory Group on Ocean Research of the World Meteorological Organization, Ponza and Rome, 29 April to 7 May 1969. La Jolla, California: 1969.

Kuhn, A. G. "NATO and Disarmament." *NATO Letter,* no. 1 (1969): pp. 17-19.

McKelvey, V. E., and Wang, Frank F. H. *World Subsea Mineral Resources: A Discussion to Accompany Miscellaneous Geologic Investigations Map 1-632.* Washington: Department of the Interior, U. S. Geological Survey, 1969.

United Nations, Economic and Social Council. *Marine Science and Technology: Survey and Proposals—Report of the Secretary General* (E/4487), 24.

United Nations, Food and Agricultural Organization. *The State of the World's Fisheries.* Rome: 1968.

United Nations, General Assembly, Secretariat. *Legal Aspects of the Question of the Reservation Exclusively for Peaceful Purposes of the Sea-Bed and the Ocean Floor, and the Subsoil thereof, Underlying the High Seas Beyond the Limits of Present National Jurisdiction, and the Use of Their Resources in the Interests of Mankind,* (A/AC.135/19 [and Add. 1 and 2]), 21 June 1968.

United Nations, Intergovernmental Oceanographic Commission. *Manual on International Oceanographic Data Exchange.* 2d ed., rev. Paris: UNESCO, 1967.

U. S., Arms Control and Disarmament Agency. *Eighth Annual Report to*

Congress: 1 January 1968 to 31 December 1968. Washington, D. C.: 1969.

U. S., Commission on Marine Science Engineering and Resources. *Panel Reports of the Commission on Marine Science, Engineering and Resources:* Volume 1–"Science and Environment"; Volume 2–"Industry and Technology: Keys to Ocean Development"; Volume 3–"Marine Resources and Legal-Political Arrangements for Their Development." Washington: Govt. Print. Off., 1969.

U. S., Congress, Senate. *Senate Resolution 33: Declaration of Legal Principles Governing Activities of States in the Exploration and Exploitation of Ocean Space.* Submitted by Senator Pell, Rhode Island, 91st Cong., 1st sess., 2 January 1969.

——, Congress, Senate. *Senate Resolution 263: Treaty on Principles Governing the Activities of States in the Explorations and Exploitation of Ocean Space.* Submitted by Senator Pell, Rhode Island. 90th Cong., 2d sess., 5 March 1968.

U. S., Department of Interior and Department of Transportation. *Oil Pollution–A Report to the President: A Special Study by the Secretary of the Interior and the Secretary of Transportation.* Washington: 1968.

U. S., Department of the Interior, National Petroleum Council (Committee on Petroleum Resources Under the Ocean Floor). *Petroleum Resources Under the Ocean Floor.* Washington: 1969.

U. S., Executive Office of the President, National Council on Marine Resources and Engineering Development. *International Decade of Ocean Exploration: A Report.* Washington: U. S. Govt. Print. Off., 1968.

——, Executive Office of the President, National Council on Marine Resources and Engineering Development. *Marine Science Activities of the Nations of Africa, East Asia, Europe, Latin America, and the Near East and South Asia.* 5v. Washington: U. S. Govt. Print. Off., 1968.

——, Executive Office of the President, National Council on Marine Resources and Engineering Development. *Marine Science Affairs: A Year of Broadened Participation.* The third report of the President to the Congress. Washington: U. S. Govt. Print. Off., 1969.

U. S., National Research Council of the National Academy of Sciences–National Academy of Engineering. *Bulletin of the Academy of Sciences, USSR, Izvestiya, Atmospheric and Oceanic Physics Series* (translation). Monthly. Washington: National Academy of Sciences.

U. S., National Research Council, National Academy of Engineering (Committee on Ocean Engineering) and the National Academy of Sciences (Committee on Oceanography). *An Oceanic Quest: The International Decade of Ocean Exploration.* (Nat. Acad. of Sciences Pub. 1709). Washington: National Academy of Sciences, 1969.

U. S., Public Land Law Review Commission. *Study of Outer Continental Shelf Lands of the United States.* 2 vols. Prepared under contract with the Public Land Law Review Commission by Nossaman, Waters, Scott, Krueger & Riordan (Robert B. Krueger, Project Director). Los Angeles: 1968.

ARTICLES & PAPERS

Andrassy, Juraj. "Epikontinentalni pojas i medunarodno obicajno pravo." *Medunarodni Problemi* 2 (1968): 216-51.

——. "Exploitation of Deep Sea Resources." Jugoslovenska *Revija za Medunarodno Pravo* 15 (1968): 98-.

Basiuk, Victor. "The Oceans and Foreign Policy: Laissez-faire or a Stronger National Purpose?" In *Law of the Sea: National Policy Recommendations,* edited by Lewis M. Alexander, pp. 71-79. Kingston, R. I.: University of Rhode Island, 1970.

Basiuk, Victor. "Marine Resources, Development, Foreign Policy and the Spectrum of Choice." *Orbis,* 12 (Spring 1968): 39-72.

Borgese, Elisabeth Mann. "A Center Report—The Republic of the Deep Seas." *The Center Magazine,* no. 4 (1968), pp. 18-27.

Bouchez, L. J., Rapporteur. *Deep-Sea Mining: Report of the Deep-Sea Mining Committee on the Exploration and Exploitation of Minerals on the Ocean Bed and its Subsoil.* International Law Association, Buenos Aires Conference, 1968.

Brooks, D. L. "Deep Sea Manganese Nodules: From Scientific Phenomenon to World Resource." *Natural Resources Journal* 8 (1968): 401-23.

Brown, E. D. "Our Nation and the Sea: A Comment on the Proposed Legal-Political Framework for the Development of Submarine Mineral Resources." In *Law of the Sea: National Policy Recommendations,* edited by Lewis M. Alexander, pp. 2-49. Kingston, R. I.: University of Rhode Island, 1970.

Browning, David S. "The United Nations and Marine Resources." *William and Mary Law Review* 10 (Spring 1969): 690-705.

Brownlie, Ian. "Recommendations on the Limits of the Continental Shelf and Related Matters." In *Law of the Sea: National Policy Recommendations,* edited by Lewis M. Alexander, pp. 133-58. Kingston, R.I.: University of Rhode Island, 1970.

Burke, William T. "Contemporary Legal Problems in Ocean Development." In *Towards the Better Use of the Oceans: A Study and Prognosis,* pp. 15-204. Stockholm: International Institute for Peace and Conflict Research, 1968.

——. "A Negative View of a Proposal for United Nations Ownership of Ocean Mineral Resources." *Natural Resources Lawyer* 1 (June 1968): 42-62.

Butler, William E. "The Soviet Union and the Continental Shelf." *American Journal of International Law* 63 (1969): 103-107.

Calder, Nigel. "Undersea Colonialism." *New Scientist* (1969): 322-32.

Carroz, J. E., and Roche, A. G. "The International Policing of High Sea Fisheries." *Canadian Yearbook of International Law* (1968): 61-90.

Chanhan, B. R. "The Position of Land-Locked States in International Law." *Law Review (Punjab [State] Law College)* 18 (1966): 422-40.

Chapman, Wilbert McLeod. "On the United States Fish Industry and the 1958 and 1960 United Nations' Conference on the Law of the Sea." In *The Law of the Sea: International Rules and Organization for the Sea,* edited by L. M. Alexander, pp. 35-63. Kingston, R.I.: University of Rhode Island, 1969.

Charles, H. "Les îles artificielles." *Revue Generale de Droit International Public* 71 (1967): 342-68.

Cheever, Daniel S. "International Organizations for Marine Science: An Eclectic Model." In *Law of the Sea: National Policy Recommendations,* edited by Lewis M. Alexander, pp. 377-90. Kingston, R.I.: University of Rhode Island, 1970.

Cheng, T. "Communist China and the Law of the Sea." *American Journal of International Law* 63 (1969): 47-73.

Cheprow. "International Regime of the Seabed (in Russian)." *The Soviet State and Law* 10 (1968).

Christy, Francis T., Jr. "Alternative Regimes for the Marine Resources Underlying the High Seas." *Natural Resources Lawyer* 1 (1968): 63-77.

——, and Scott, Anthony. "The Common Wealth in Ocean Fisheries: Some Problems of Growth and Economic Allocation." (Published for Resources for the Future.) Baltimore: Johns Hopkins Press, 1965.

——. "Mineral Resources of the Sea-Bed Other Than Petroleum and Natural Gas: Marigenous Minerals—Wealth, Regimes and Factors of Decision." Paper presented at Symposium on the International Regime of the Sea-Bed, 30 June to 5 July 1969, Rome. Mimeographed.

Cloud, Preston. "Mineral Resources from the Sea." In *Resources and Man.* Washington: National Academy of Sciences, Committee on Resources and Man, 1969.

Danzig, A. L. "Proposed Treaty Governing the Exploration and Use of the Ocean Bed." New York: United Nations Committee of the World Peace Through Law Center, 1968.

"Deep Ocean Oil Prospects Declared Dim." *Oil and Gas Journal,* 16 December 1968, pp. 50-51.

de Sylva, Donald P. "The Unseen Problems of Thermal Pollution." *Oceans Magazine* 1 (January 1969): 38-41.

Eichelberger, Clark M. "The United Nations and the Bed of the Sea." *San Diego Law Review,* no. 3 (1969).

Ely, Northcutt. "A Case for the Administration of Mineral Resources Underlying the High Seas by National Interests." *Natural Resources Lawyer* 1 (June 1968): 78-84.

Emery, K. O. "An Oceanographer's View of the Law of the Sea." In *Law of the Sea: National Policy Recommendations,* edited by Lewis M. Alexander, pp. 211-25. Kingston, R.I.: University of Rhode Island, 1970.

Friedmann, W. "The Race to the Bottom of the Sea." *Columbia Forum,* no. 1 (1969), pp. 18-21.

Friedrich, Hermann. "Food from the Sea." *Universitas* 11 (1968): 171-77.

Goldie, L. F. C. "Submarine Zones of Special Jurisdiction Under the High Seas: Some Military Aspects." In *The Law of the Sea: The Future of the Sea's Resources,* edited by Lewis M. Alexander, pp. 100-13. Columbus, Ohio: Ohio State University Press, 1967.

——. "Sedentary Fisheries and Art. 2(4) of the Continental Shelf Convention: A Plea for a Separate Regime." *American Journal of International Law* 63 (1969): 86-97.

Gordon, H. Scott. "The Economic Theory of a Common Property Resource: The Fishery." *The Journal of Political Economy* 62 (1954): 124-42.

Haight, G. W. "United Nations Affairs: Ad Hoc Committee on Sea-Bed and Ocean Floor." *The International Lawyer* 3 (1968): 22-30.

Hearn, Wilfred A. "The Role of the United States Navy in the Formulation of Federal Policy Regarding the Sea." *Natural Resources Lawyer* 1 (June 1969): 23-31.

Hersh, S. M. "An Arms Race on the Sea Bed?" *War/Peace Report,* no. 7 (1968), pp. 8-9.

International Institute of Peace and Conflict Research. *Towards the Better Use of the Oceans: A Study and Prognosis.* Stockholm: 1968.

International Law Association. *Interim Report of the Committee on Deep Sea Minerals of the American Branch of the International Law Association,* 19 July 1968.

Kalinkin, G. "Military Use of the Sea-Bed Should Be Banned." *International Affairs* (Moscow), no. 2 (1969), pp. 45-48.

Kasahara, Hiroshi. "Food Production from the Ocean." In *Proceedings of the Conference on Law, Organization and Security in the Use of the Ocean at Columbus, Ohio, 1967.*

Kehden, M. J. "Die Vereinten Nationen und die Nutzung des Bodens und Untergrundes des Hohen Meeres ausserhalb der Grenzen nationaler Hoheitsgewalt." *Verfassung und Recht in Ubersee* 2 (1969): 131-67.

Kiefe, R. "Les aspects juridiques des pollutions marine." *Revue International D'Oceanographie Medicale* 11 (1968): 187-91.

Klima, Otto, Jr., and Wolfe, Gibson M. "The Oceans: Organizing for Action." *Harvard Business Review,* May–June 1968, pp. 98-112.

——. "The Oceans: Unexploited Opportunities." *Harvard Business Review,* March–April 1968, pp. 140-156.

Krueger, Robert B. "The Convention of the Continental Shelf and the Need for Its Revision and Some Comments Regarding the Regime for the Lands Beyond." *Natural Resources Lawyer* 1 (July 1968): 1-18.

Langeraar, W. "Some Thoughts on an International Regime and Administrating Agency for the Seabed and Ocean Floor Beyond the Limits of National Jurisdiction." In *Law of the Sea: National Policy Recommendations,* edited by Lewis M. Alexander, pp. 110-19. Kingston, R.I.: University of Rhode Island, 1970.

Manheim, Frank T. "Soviet Books in Oceanography." *Science* 154 (1966): 995-98.

Marine Biological Association of the United Kingdom (Plymouth Laboratory). *'Torrey Canyon' Pollution and Marine Life: A Report by the Plymouth Laboratory of the Marine Biological Association of the United Kingdom.* Edited by J. E. Smith. Cambridge: University Press, 1968.

McKernan, Donald L. "International Fishery Regimes: Current and Future." In *Law of the Sea: National Policy Recommendations,* edited by Lewis M. Alexander, pp. 336-44. Kingston, R.I.: University of Rhode Island, 1970.

Miron, George. "Proposed Regimes for Exploration and Exploitation of the Deep-Seabed." In *Law of the Sea: National Policy Recommendations,* edited by Lewis M. Alexander, pp. 98-109. Kingston, R.I.: University of Rhode Island, 1970.

Misra, K. P. "Territorial Sea and the India." *Indian Journal of International Law* 6 (1966): 465-82.

Neild, R. R. "Alternative Forms of International Regime for the Oceans." In *Towards a Better Use of the Oceans: A Study and Prognosis,* pp. 279-92. Stockholm: International Institute for Peace and Conflict Research, 1968.

Nierenberg, William H. "Militarized Oceans." In *Unless Peace Comes: A Scientific Forecast of New Weapons,* edited by Nigel Calder, pp. 109-19. London, 1968.

Oda, Shigeru. "Boundary of the Continental Shelf." *Japanese Annual of International Law* 12 (1968): 264-84.

Pardo, Arvid. "Sovereignty Under the Sea: The Threat of National Occupation." *Round-Table* 232 (October 1968) 341-55.

——. "Who Will Control the Seabed?" *Foreign Affairs* 45 (1968): 123-37.

——. "Whose Is The Bed of the Sea?" *American Society of International Law Proceedings* (1968), pp. 216-29.

——, and Gauchi, V. "The Sea Bed: Common Heritage of Mankind." *War/Peace Report,* no. 7 (1968), pp. 3-6.

Rao, P. Sreenivasa. "The Law of the Continental Shelf." *Indian Journal of International Law,* 1966.

Rich, A., and Engelhardt, V. A. "A Proposal from A U. S. and A Soviet Scientist: Oceanic Resources and Developing Nations." *Bulletin of the Atomic Scientist,* 1968.

Ricker, William E. "Food from the Sea." In *Resources and Man.* Washington: National Academy of Sciences, Committee on Resources and Man, 1969.

Stephanova, S. "The Legal Classification of Ocean Space." In Russian. *Grdischnik na Sofijskija Universitet* 55 (1964): 111-205.

Taube, G. "Militar und Meer—Das Ringen um die Vorherrschaft in der Tiefsee." *Wehrkunde* (1967): 492-94.

Thomer, E. "U-Boote in Ost und West (Submarines in the East and West)." *Wehrkunde* 17 (1968): 212-16.

Volkov, A. A. "Contemporary Principles of International Regulation of Open Sea Fishing." In Russian. *Soviet Yearbook of International Law,* 1966-1967 (1968): 203-18.

Weissberg, Guenter. "International Law Meets the Short-Term National Interest: The Maltese Proposal on the Sea-Bed and Ocean Floor—Its Fate in Two Cities." *International and Comparative Law Quarterly* 18 (1969): 41-102.

Wenk, Edward, Jr. "A New National Policy for Marine Resources." *Natural Resources Lawyer* 1 (June 1968): 3-13.

Werness, M. W. "The Eighth Ocean." *U. S. Anti-Submarine Warfare Quarterly* (1968): 14-15.

Wilkey, M. R. "The Deep Sea: Its Potential Mineral Resources and Problems." *The International Lawyer* 3 (1968): 31-48.

Young, Richard. "The Legal Regime of the Deep-Sea Floor." *American Journal of International Law* 62 (1968): 641-53.

Note. "In Re Reference Concerning Ownership of and Jurisdiction over Offshore Mineral Rights. *Ottawa Law Review* 2 (1967): 212-.

PROCEEDINGS AND PERIODICALS

Akademia nauk U.S.S.R. Institut okeanologii. *The Pacific Ocean.* Edited by V. G. Kort. Moscow: Nauka, 1966-. [Alexandria, Va.: Clearinghouse for Federal Scientific and Technical Information, 1966-.] Translation prepared for U. S. Naval Oceanographic Office.

The American Assembly, Columbia University. *Uses of the Seas: Intended as Background Reading for the 33d American Assembly, Arden House, May 2-5, 1968.* Edited by Edmund A. Gullion. Englewood Cliffs, N. J.: Prentice-Hall, 1968.

Conference on the Technology of the Sea and the Sea-Bed. *Proceedings of the Conference . . . held at the Atomic Energy Research Establishment, Harwell, April 5th, 6th and 7th, 1967: Sponsored by the Ministry of Technology.* London, H. M. S. O., 1967.

Federazione delle associazioni scientifiche e techniche and others. *Symposium on Fresh Water from the Sea. Acqua dolce del mare. Il inchiesta internazionale.* Atti del simposio internazionale tenutosi a Milano per iniziativa del Gruppo di studio della acque della Federazione della associazioni scientifiche e techiche, in collaborazione con l'Ente autonomo Fiera di Milano e il consiglio nazionale delle ricerche. Roma: Consiglio nazionale delle ricerche, 1967.

International Conference on Oil Pollution of the Sea, Rome. 7-9 October 1968. *Report of Proceedings.* Sponsored by The British Advisory Committee on Oil Pollution of the Sea; The Italian National Committee on Oil Pollution of the Sea; and the Nordic Union for the Prevention of Oil Pollution of the Sea. Winchester, England: Warren and Son, Ltd., The Wykeham Press, 1968.

Oceanography and Marine Biology: An Annual Review. Edited by H. Barnes. v. 1-. London: Allen & Unwin, 1965.

Oceanographie physique. Paris: Centre national de la recherche scientifique, 1962-.

Oceanology of the Academy of Sciences of the USSR. (successor to *Soviet Oceanography*). Translated and edited by Scripta Technica, Inc. (Six numbers annually). Washington: American Geophysical Union: 1965-.

BOOKS AND MONOGRAPHS

Alexander, Lewis M., ed. *The Law of the Sea: The Future of the Sea's Resources.* Kingston: University of Rhode Island, 1968.

———. *The Law of the Sea: International Rules and Organization for the Sea.* Kingston: University of Rhode Island, 1969.

———. *The Law of the Sea: Offshore Boundaries and Zones.* Columbus: Ohio State University Press, 1967.

Alvarado Garaicoa, Teodoro. *El domino del mar.* Guayaquil: Departamento de Publicaciones de la Universidad de Guayaquil, 1968.

Bohm, Eckart. *Tankerunfalle auf dem Hohen Meer: Die Zulassigkeit staatlicher Mabnahmen zur Gefahrenabwehr.* Hamburg: Forschungsstell fur Volkerrecht und auslandisches offentliches Recht der Universitat Hamburg, 1970.

Borgese, Elisabeth Mann. *The Ocean Regime: A Suggested Statute for the Peaceful Uses of the High Seas and the Sea-Bed Beyond the Limits of National Jurisdiction.* A Center Occasional Paper. Santa Barbara, California: Center for the Study of Democratic Institutions, 1968.

Brahtz, John F., ed. *Ocean Engineering: Goals, Environment, Technology.* New York: Wiley, 1968.

Butler, William E. *The Law of Soviet Territorial Waters: A Case Study of Maritime Legislation and Practice.* New York: Praeger, 1967.

Calder, Nigel, ed. *The World in 1984: The Complete "New Scientist" Series.* 2 vols. Harmondsworth: Penguin, 1965.

Christy, Francis T., and Scott, Anthony. *The Common Wealth in Ocean Fisheries: Some Problems of Growth and Economic Allocation.* Published for Resources for the Future. Baltimore: Johns Hopkins Press, 1965.

Colombos, Constantine John. *The International Law of the Sea.* 6th rev. ed. London: Longmans, 1967.

Cousteau, Jacques-Yves. *World Without Sun.* Edited by James Dugan. London: Heinemann, 1965. [New York: Harper & Row, 1965.]

Crutchfield, James Arthur, and Pontecorvo, Giulio. *The Pacific Salmon Fisheries: A Study of Irrational Conservation.* Published for Resources for the Future. Baltimore: Johns Hopkins Press, 1969.

——. *Physical Oceanography.* 2 vols. Oxford and New York: Pergamon Press, 1961.

Fattal, Antoine. *Les conferences des Nations Unies et la Convention de Genève du 29 avril 1958 sur la mer territoriale et la zone contigue.* Beyrouth: Librairie du Liban, 1968.

Fraser, James. *Treibende Welt: eine Naturgeschichte des Meersplanktons.* Berlin, New York: Springer-Verlag, 1965.

Friedheim, Robert L. *Understanding the Debate on Ocean Resources.* Law of the Sea Institute Occasional Paper No. 1. Kingston, R. I.: University of Rhode Island, 1968.

Gretton, Sir P. *Maritime Strategy: A Study of Defense Problems.* New York: Praeger, 1965.

Groen, Pier. *The Waters of the Sea.* London, Princeton: Van Nostrand, 1967.

Henkin, Louis. *Law for the Sea's Mineral Resources.* New York: Institute for the Study of Science in Human Affairs, Columbia University, c. 1968.

Hickling, Charles Frederick. *The Farming of Fish.* Oxford, New York: Pergamon Press, 1968.

Iverson, E. S. *Farming the Edge of the Sea.* London: Fishing News (Books) Ltd., 1968.

Kuenen, Phillip Henry. *Marine Geology.* New York: Wiley, 1963.

Martin, Laurance W. *The Sea in Modern Strategy.* London: Chatto & Windus for the Institute for Strategic Studies, 1967.

Matte, Nicolas Mateesco. *Deux frontières invisibles: De le mer territoriale à l'air "territorial."* Paris: Editions A Pedone, 1965.

Mateesco, Mircea. *Le droit maritime et la droit aérien de l'U.R.S.S. a l'hure de la coexistence pacifique.* Paris: Editions A Pedone, 1967.

Melesio Montoya, Octavio. *El mar territorial y la llamada zone contigua en el derecho international.* Mexico: 1964.

Mero, John L. *The Mineral Resources of the Sea.* Amsterdam, New York: Elsevier Pub. Co., 1965.

Merryman, J. H., and Ackerman, E. D. *International Law, Development and the Transit Trade of Landlocked States: The Case of Bolivia.* Hamburg: Forschungsstelle für Völkerrecht und ausländisches öffentliches Recht der Universität Hamburg, 1969.

Neukirchen, H. *Krieg zur Sea (War at Sea).* Berlin: Deutscher Militärverlag, 1966.

The Ocean. (Ten articles from the September 1969 issue of the *Scientific American.*) San Francisco: W. H. Freeman, 1969.

Oceanic Research Institute. *Oceanic Index.* v. 1-. La Jolla, Calif.: Oceanic Library and Information Center, 1964-.

Oda, Shigeru. *International Control of Sea Resources.* Leyden: A. W. Sythoff, 1963.

Pell, Claiborne, and Goodwin, Harold Leland. *Challenge of the Seven Seas.* New York: Morrow, 1966.

Percier, Albert. *Océanographie et technique des pêches maritimes: Cours professé à l'École nationale des affaires maritimes, Bordeaux.* Biarritz: Centre d'étude et de recherches scientifiques, 1967.

Polikarpov, Gennadii Grigor'evich. *Radioecology of Aquatic Organisms: The Accumulation and Biological Effect of Radioactive Substances.* Translated from the Russian by Scripta Technica, Ltd. Edited by Vincent Schultz and Alfred W. Clement, Jr. New York: Reinhold Book Division, 1966.

Reed, Laurance. *Ocean-Space–Europe's New Frontier: Towards a Long-Range, Concerted Programme for Exploiting the Resources of the Sea.* London: Bow Publications, 1969.

Shepard, Francis P. *The Earth Beneath the Sea.* Rev. ed. Baltimore: Johns Hopkins Press, 1967. [New York: Atheneum, 1964.]

Werner, Auguste R. *Traite de droit maritime general: Elements et systeme, definitions, problemes, principes.* Geneve: Librairie Droz, 1964.

World Peace Through Law Center, United Nations Committee. *Treaty Governing the Exploration and Use of the Ocean Bed.* Its Pamphlet series, 10. Geneva: World Peace Through Law Center, 1968?

Yoshida, Kozo, ed. *Studies on Oceanography: A Collection of Papers Dedicated to Koji Hidaka in Commemoration of His Sixtieth Birthday.* American ed. Seattle: University of Washington Press, 1965.

Zaorski, Remigu Remigiusz. *Eksploatacja biologicznych zasobow morza w swietle prawa meidzynarodowego.* Gdynia, Wsyd. Morskie: 1967.